Time's Arrow &

Archimedes' Point

Time's Arrow &
Archimedes' Point

NEW DIRECTIONS FOR THE PHYSICS OF TIME

Huw Price

OXFORD UNIVERSITY PRESS

New York Oxford

Oxford University Press

Oxford   New York
Athens   Auckland   Bangkok   Bogotá   Bombay
Buenos Aires   Calcutta   Cape Town   Dar es Salaam
Delhi   Florence   Hong Kong   Istanbul   Karachi
Kuala Lumpur   Madras   Madrid   Melbourne
Mexico City   Nairobi   Paris   Singapore
Taipei   Tokyo   Toronto   Warsaw

and associated companies in
Berlin   Ibadan

Library of Congress Cataloging-in-Publication Data
Price, Huw, 1953–
Times's arrow and Archimedes' point:
new directions for the physics of time/Huw Price.
p. cm. Includes bibliographical references and index.
ISBN 0-19-510095-6
ISBN 0-19-511798-0 (Pbk.)
1. Time. 2. Physics—Philosophy. I. Title.
BD638.P73 1996 95-25508
523.1—dc20

3 5 7 9 10 8 6 4 2

Printed in the United States of America
on acid-free paper

For AHR and SMD

Preface

Time flies like an arrow; fruit flies like a banana.

MARX.

SCIENCE, like comedy, often demands that we look at familiar things in unfamiliar ways. Miss the new angles, and we miss the point. In comedy it is the comic's job to pitch the task at the right level. Too low, and the joke isn't funny; too high, and the audience doesn't get it. In science, of course, we are on our own. There are no guarantees that Nature's gags have been pitched within reach. Great scientists spend lifetimes trying to nut out the hard ones.

This book is about one of these perspective shifts—about the need to look at a familiar subject matter from a new vantage point. The subject matter concerned is one of the most familiar of all: it is time, and especially the direction of time. Despite its familiarity, time remains profoundly puzzling. It puzzles contemporary physicists and philosophers who spend large amounts of it thinking about it, as well as countless reflective nonspecialists, in search of a deeper understanding of one of the most central aspects of human life.

This book is about the need to think about time's puzzles from a new viewpoint, a viewpoint *outside* time. One of my main themes is that physicists and philosophers tend to think about time from too close up. We ourselves are creatures in time, and this is reflected in many ordinary ways of thinking and talking about the world. This makes it very difficult to think about time in an objective way, because it is always difficult to tell whether what we think we see is just a product of our vantage point. In effect, we are too close to the subject matter to see it objectively, and need to step back.

This a familiar idea in the history of science. For example, it took our ancestors a long time to figure out that the Earth and a pebble are the same kind of thing, differing only in size. To take this revolutionary idea on board, one needs to imagine a vantage point from which the Earth and the pebble can both be seen for what they are. Archimedes went one better, and offered to move the Earth, if someone would supply him with this vantage point, and a suitable lever.

I want to show that a temporal version of this Archimedean vantage point provides important insights into some of the old puzzles about time. One of its most useful roles is to highlight some old and persistent mistakes that physicists tend to make when they think about the direction of time. More surprisingly, this viewpoint also has important benefits elsewhere in physics. In particular, it throws some fascinating new light on the bizarre puzzles of quantum mechanics. Thus the book offers a novel approach to some of the most engaging issues in contemporary physics, as well as a new perspective on some of the familiar puzzles of time itself.

The book is addressed to interested nonspecialists, as well as to physicists and philosophers. In part, this is a kind of fortunate accident. My original thought was to try to make the book accessible to physicists as well as to philosophers (my home team). Many of its conclusions were going to be aimed at physicists, and I realized that there was no point in writing a book that much of the intended audience could not understand. At the same time, however, I wanted the book to be interesting and useful to my philosophical colleagues and students, most of whom have no training in physics. So I aimed for a book which would be accessible to physicists with no training in philosophy and to philosophers with no training in physics. The happy result, I think, is a book which will interest many people whose formal education qualifies on both counts: no philosophy *and* no physics.

I've been thinking about these things for a long time. As an undergraduate at ANU, Canberra, in the mid-1970s, the philosophy of time played a large part in my decision to abandon mathematics for philosophy. (I had the good fortune to encounter, in person, the very different perspectives on time of Genevieve Lloyd and Hugh Mellor.) I was an almost instant convert to the atemporal "block universe" view of time (described in chapter 1), at least for the purposes of physics. This view remains the key to the argument of the whole book.

A couple of years after that, I was already thinking about some of the issues about physics that turn up later in the book. I remember listening to a discussion of Bell's Theorem and quantum mechanics at a philosophy seminar in Oxford, and being struck by the thought that one of its crucial assumptions was time-asymmetric, in a way which looks very odd from the kind of atemporal viewpoint that goes with the block universe view. I think that I was right, but the issue turned out to be much more complicated than I then imagined, and it has taken a long time to disentangle all the threads. Strangely, one of the crucial threads goes back to the work of Michael Dummett, the Oxford philosopher who was the speaker that day in 1977—though his topic had nothing to do with the relevant part of his earlier work, as far as I can recall.

A couple of years later again, now a graduate student in Cambridge, I learned more about the physics of time asymmetry. One wet weekend in the spring of 1979, I attended a small conference on the philosophy of time in Barnstable, Devon. One of the invited speakers was Paul Davies, then a young lecturer in theoretical physics at King's College, London, who talked about the latest ideas on time in cosmology. I remember asking him afterwards why cosmologists continued to take for granted that the present state of the universe should be explained in terms of its earlier state, rather than its later state. From the atemporal perspective, I felt, this temporal bias looked rather puzzling. I can't remember exactly what Davies said in reply, but I am sure I failed to convince him that there was anything suspicious going on. But I think the failing wasn't entirely mine: I have learned that even at this level, it isn't unusual for physicists and philosophers to have trouble seeing the in-built temporal asymmetries in the ways we think about the world.

After graduate school, other philosophical projects kept me busy, and for several years I had little time for time. In 1988–1989, however, with another book finished, and a new job in the Research School of Social Sciences at ANU, I was able to pick up the threads. I became increasingly convinced that physicists tended to make serious mistakes when they thought about time, and especially about the direction of time—the kind of mistakes that careful philosophical thought could help to set right. And the underlying cause of most of these mistakes, I felt, was a failure to look at the problems from a sufficiently distant vantage point. Thus the basic project of the book was laid down.

I moved to the University of Sydney in the (southern) winter of 1989. Since then, in trying to extract the book from the gaps between other projects and responsibilities, I have been much assisted by research funding from the University's Research Grant Scheme (1991) and the Australian Research Council (1992–1993). I have also learned a lot from my students. For several years I have tried out these ideas on mixed classes of advanced undergraduates in philosophy and physics. Their reactions and comments—especially those of the rather skeptical physicists—have been invaluable in helping me to clarify my views. Among my graduate students, I am grateful to Phillip Hart, Nicholas Smith, and Patrick Yong for their comments, criticism and encouragement; and especially to Phil Dowe, now a colleague, with whom I have had many useful discussions about causal asymmetry and other things.

In the course of this long project, many other people have helped me with comments on drafts, or discussions or correspondence on particular topics. I am variously indebted to David Albert, John Baez, John Bell, Jeremy Butterfield, Craig Callender, David Chalmers, Paul Davies, Jan Faye, John Gribbin, Dan Hausman, Paul Horwich, Raymond Laflamme, Stephen Leeds, John

Leslie, David Lewis, Storrs McCall, Peter Menzies, Graham Nerlich, Graham Oppy, David Papineau, Roger Penrose, Daniel Quesada, Steve Savitt, Jack Smart, Jason Twamley, Robert Weingard, and Dieter Zeh—and, I suspect and fear, to many others whose specific contributions I cannot now recall.

Two of these people deserve special mention. Jack Smart is an Australian philosopher, well known, among other things, for his work on the philosophy of time. (Twenty years ago, when I first encountered the subject, his work was already classic.) Because he is an exponent of the block universe view, as well as a generous and enthusiastic man, I expected him to be positive about the early drafts of this book. Even so, the warmth of his response has surprised me, and his comments and enthusiasm have been a very great source of encouragement.

Dieter Zeh, of Heidelberg University, is well known among physicists for his work on the direction of time. He wrote to me in 1989, responding to an article which had just appeared in *Nature,* in which I criticized some of Stephen Hawking's claims about the direction of time. I felt rather hesitant about taking on such a famous opponent in such a public forum, so it was a great relief and encouragement when Zeh's note arrived, saying "I agree with every word you say about Hawking." We have been regular correspondents since then, and although there are many points on which we continue to disagree, these exchanges have been an important source of insight and encouragement, as the book has come together.

Some of the book draws on material I have previously published elsewhere. Chapter 3 relies heavily on the article listed in the Bibliography as Price (1991c), chapter 4 on (1995), chapter 6 on (1992a), and parts of chapters 7 and 9 on (1994). I am grateful to the editors and publishers concerned for permission to reuse the material in this form.

Finally, two more personal acknowledgments. I am very warmly indebted to Nye Rozea, not least for his cheerful and unflagging skepticism about the entire project—indeed, about my intellectual capacities in general. This proved a priceless antidote to self-esteem, and I'm not sure which of us will be more surprised to see the book finished. Nye's generous filial skepticism was tempered, happily, by the support and enthusiasm—more considered, I think, but perhaps therefore even more generous—of Susan Dodds. To these two friends, then, for what it's worth: take this …

Contents

Time's Arrow &
Archimedes' Point

1

←——————→

The View from Nowhen

S<small>AINT</small> A<small>UGUSTINE</small> (354–430) remarks that time is at once familiar and deeply mysterious. "What is time?" he asks. "If nobody asks me, I know; but if I were desirous to explain it to one that should ask me, plainly I know not."[1] Despite some notable advances in science and philosophy since the late fourth century, time has retained this unusual dual character. Many of the questions that contemporary physicists and philosophers ask about time are still couched in such everyday terms as to be readily comprehensible not only to specialists on both sides of the widening gulf between the two subjects— that in itself is remarkable enough—but also to educated people who know almost nothing about either field. Time is something rather special, then. Few deep issues lie so close to the surface, and fewer still are yet to be claimed by a single academic discipline.

This book is concerned with a particular kind of question about time. What is the difference between the past and the future? Could—and does— the future affect the past? What gives time its direction, or "arrow"? Could time be symmetric, or a universe be symmetric in time? What would such a world be like? Is our world like that? The book is concerned with what modern physics has to say about issues of this kind, but I am not writing as a physicist, explaining the insights of my discipline to a general audience. I am a philosopher, and the vantage point of the book is philosophical. One of my main aims is to sort out some philosophical confusions in the answers that contemporary physicists typically give to these questions. I want to provide physicists themselves, as well as philosophers and general readers, with a clearer picture of these issues than has yet been available.

What are these philosophical confusions? The most basic mistake, I shall be arguing, is that people who think about these problems—philosophers as well as physicists—often fail to pay adequate attention to the temporal

3

character of the viewpoint which we humans have on the world. We are creatures *in* time, and this has a very great effect on how we think *about* time and the temporal aspects of reality. But here, as elsewhere, it is very difficult to distinguish what is genuinely an aspect of reality from what is a kind of appearance, or artifact, of the particular perspective from which we regard reality. I want to show that a distinction of this kind is crucial to the project of understanding the asymmetry of time. In philosophy and in physics, theorists make mistakes which can be traced to a failure to draw the distinction sufficiently clearly.

The need to guard against anthropocentrism of this kind is a familiar theme in the history of both science and philosophy. One of the great projects in the history of modern thought has been the attempt to achieve the untainted perspective, the Archimedean view of reality—"the view from nowhere," as the philosopher Thomas Nagel calls it.[2] The main theme of this book is that neither physics nor philosophy has yet paid enough attention to the temporal aspect of this ancient quest. In particular, I want to show that if we want to understand the asymmetry of time then we need to be able to understand, and quarantine, the various ways in which our patterns of thought reflect the peculiarities of our own temporal perspective. We need to acquaint ourselves with what might aptly be called the view from *nowhen*.

Our interest in questions of temporal asymmetry thus lies at more than one level. There is the intrinsic interest of the physical issues themselves, of course, and the book aims to present a clearer, more insightful, and more accessible view of the main problems and their possible resolutions than has yet been available. In criticizing previous writers, however, my main argument will be that when discussing temporal asymmetry, they have often failed to disentangle the human temporal perspective from the intended subject matter. And it is the asymmetry of our ordinary temporal perspective which is the source of the difficulty, so that the task of unraveling the anthropocentric products of this perspective goes hand in hand with that of deciding how much of temporal asymmetry is really objective, and therefore in need of explanation by physics.

The book thus straddles the territory between physics and philosophy. On the physical side, my main goal will be to obtain a clear view of the problem, or problems, of the asymmetry of time, to correct certain common errors in existing approaches to the problem, and to assess current prospects for a solution. But the main contribution I bring to these problems will be a philosophical one, particularly that of showing how errors arise from a failure to distinguish between the viewpoint we have from within time and the Archimedean standpoint from which physics needs to address these issues. On the purely philosophical side, I shall be interested in the project of characterizing this view from nowhen—of deciding which features of the ordinary world remain visible from this perspective, for example, and which turn out to depend on the temporal viewpoint we normally occupy.

Perspective shifts of this kind are nothing new in science, of course. Some of the most dramatic revolutions in the history of science have been those that have overturned previous conceptions of our own place in nature. The effect is something like that of coming suddenly to a high vantage point—at once exciting and terrifying, as a familiar view of our surroundings is revealed to be a limited and self-centered perspective on a larger but more impersonal reality. In physics the most dramatic example is the Copernican revolution, with its overthrow of the geocentric view of the universe. In biology it is Darwinism, with its implications for the place of humanity in nature. These two examples are linked in the more gradual but almost equally revolutionary discovery of cosmological time (and hence of the insignificance of human history on the cosmological scale).

While the perspective shift I shall be recommending in this book is not in this league—it would be difficult to significantly dehumanize a world in which the place of humanity is already so insignificant—it does have some of their horizon-extending impact. For it turns on the realization that our present view of time and the temporal structure of the world is still constrained and distorted by the contingencies of our viewpoint. Where time itself is concerned, I claim, we haven't yet managed to tease apart what Wilfred Sellars calls the scientific and manifest images—to distinguish how the world *actually is,* from how it *seems to be* from our particular standpoint.

As in earlier cases, the intellectual constraint is largely self-imposed. To notice the new standpoint is to be free to take it up, at least for the purposes of physics. (We can't actually stand outside time, but we can imagine the physics of a creature who could.) Again the discovery is both exciting and unsettling, however, in showing us a less anthropocentric, more objective, but even more impersonal world.

OUTLINE OF THE BOOK

The remainder of this introductory chapter deals with some important preliminaries. One of these is to set aside certain philosophical issues about time which won't be dealt with later in the book. Philosophical discussions of time have often focused on two main issues, that of the objectivity or otherwise of the past-present-future distinction, and that of the status of the flow of time. Philosophers have tended to divide into two camps on these issues. On the one side are those who treat flow and the present as objective features of the world; on the other, those who argue that these things are mere artifacts of our subjective perspective on the world. For most of the book I shall be taking the latter view for granted. (Indeed, I take the central philosophical project of the book to be continuous with that of philosophers such as D. C. Williams, J. J. C. Smart, A. Grünbaum, and D. H. Mellor.)[3] I shall presuppose that we have learnt from this tradition that many of our ordinary temporal notions are anthropocentric in this way. My aim is to extend these insights, and apply them to physics. I shall not defend this presupposition in the sort of detail it receives elsewhere in the philosophical literature—that would take a book to itself—but I set out below what I see as the main points in its favor.

The second important preliminary task is to clarify what is meant by the asymmetry or arrow of time. A significant source of confusion in contemporary work on these topics is that a number of distinct notions and questions are not properly distinguished. It will be important to say in advance what our project is, and to set other issues to one side. Again, however, I shall draw these distinctions rather quickly, with no claim to be philosophically comprehensive, in order to be able to get on with the main project.

With the preliminaries out of the way, the remainder of the book is in two main parts. The first part (chapters 2–4) focuses on the three main areas in which temporal asymmetry turns up in modern physics: in thermodynamics, in phenomena involving radiation, and in cosmology. In all these cases, what is puzzling is why the physical world should be asymmetric in time at all, given that the underlying physical laws seem to be very largely symmetric. These chapters look at some of the attempts that physicists have made to solve this puzzle, and draw attention to some characteristic confusions and fallacies that these attempts tend to involve.

Chapter 2 deals with thermodynamics. Few ideas in modern physics have had as much impact on popular imagination and culture as the second law of thermodynamics. As everyone knows, this is a time-asymmetric principle. It says that entropy *increases* over time. In the late nineteenth century, as thermodynamics came to be addressed in terms of the symmetric framework of statistical mechanics, the puzzle just described came slowly into view: where does the asymmetry of the second law come from? I shall explain how, as this problem came into view, it produced the first examples of a kind of fallacy which has often characterized attempts to explain temporal asymmetry in physics. This fallacy involves a kind of special pleading, or double standard. It takes an argument which could be used equally well in either temporal direction and applies it selectively, in one direction but not the other. Not surprisingly, this biased procedure leads to asymmetric conclusions. Without a justification for the bias, however, these conclusions tell us nothing about the origins of the real asymmetry we find in the world.

Fallacies of this kind crop up time and time again. One of the main themes of this book is that we need the right starting point in order to avoid them. In chapter 2 I'll use examples from the history of thermodynamics to illustrate this idea. I shall also describe an exceptional early example of the required atemporal viewpoint, in the work of Ludwig Boltzmann, the Austrian physicist who was responsible for some of the fundamental results of the period. As we'll see, Boltzmann was perhaps the first person to appreciate the true importance of the question: Why was entropy low in the past? The chapter concludes with a discussion as to what it is that really needs to be explained about the asymmetry of thermodynamics—I shall argue that very few writers have drawn the right lesson from the nineteenth century debate—and offers some guidelines for avoiding the kinds of mistakes that have plagued this field for 150 years.

Chapter 3 looks at the time asymmetry of a wide range of physical phenomena involving radiation. Why do ripples on a water surface spread outwards rather than inwards, for example? Similar things happen with other kinds of radiation, such as light, and physicists have been puzzled by the temporal asymmetry of these phenomena since the early years of the twentieth century. In discussing this issue, it turns out to be important to correct some confusions about what this asymmetry actually involves. However, the chapter's main focus will be the issue of the relation between this asymmetry and that of thermodynamics. I want to show that several prominent attempts to reduce the former asymmetry to the latter turn out to be fallacious, once the nature of the thermodynamic asymmetry is properly appreciated. In particular, I want to look at a famous proposal by the American physicists John Wheeler and Richard Feynman, called the Absorber Theory of Radiation. At first sight, this theory seems to involve the very model of respect for an atemporal perspective. I shall show that Wheeler and Feynman's reasoning is confused, however, and that as it stands, their theory doesn't succeed in explaining the asymmetry of radiation in terms of that of thermodynamics. However, the mathematical core of the theory can be reinterpreted so that it does show—as Wheeler and Feynman believed, but in a different way—that radiation is not intrinsically asymmetric; and that its apparent asymmetry may be traced, if not to the thermodynamic asymmetry itself, then to essentially the same source. (In effect, then, I want to show that Wheeler and Feynman produced the right theory, but tried to use it in the wrong way.)

Chapter 4 turns to cosmology. As chapter 2 makes clear, the search for an explanation of temporal asymmetry leads to the question why the universe was in a very special condition early in its history—why entropy is low near the big bang. But in trying to explain why the universe is like this, contemporary cosmologists often fall for the same kind of fallacies of special pleading, the same application of a double standard with respect to the past and the future, as their colleagues elsewhere in physics. In failing to adopt a sufficiently atemporal viewpoint, then, cosmologists have failed to appreciate how difficult it is to show that the universe must be in the required condition at the big bang, without also showing that it must be in the same condition at the big crunch (so that the ordinary temporal asymmetries would be reversed as the universe recollapsed). Cosmologists who do consider the latter possibility often reject it on grounds which, if applied consistently, would also rule out a low-entropy big bang. As we shall see, the mistakes made here are very much like those made a century earlier, in the attempt to put the asymmetry of thermodynamics on firm statistical foundations. My concern in this chapter is to draw attention to these mistakes, to lay down some guidelines for avoiding them, and to assess the current prospects for a cosmological explanation of temporal asymmetry.

In the first part of the book, then, the basic project is to try to clarify what modern physics tells us about the ways in which the world turns out to be asymmetric in time, what it tells us about how and why the future is different from the past. And the basic strategy is to look at the problem from a sufficiently detached standpoint, so that we don't get misled by the temporal asymmetries of our own natures and ways of thinking. In this way, I argue, it is possible to avoid some of the mistakes which have been common in this branch of physics for more than a century.

In the second part of the book, I turn from the physics of time asymmetry to physics more generally. The big project of this part of the book is to show that the atemporal Archimedean perspective has important ramifications for the most puzzling puzzle of all in contemporary physics: the meaning of quantum theory. My view is that the most promising understanding of quantum theory has been almost entirely overlooked, because physicists and philosophers have not noticed the way in which our ordinary view of the world is a product of our asymmetric standpoint. Once we do notice it—and once we think about what kind of world we might expect, given what we have discovered about the physical origins of time asymmetry—we find that we have good reason to expect the very kind of phenomena which make quantum theory so puzzling. Quantum theory turns out to be the kind of microphysics we might have expected, in other words, given our present understanding of the physical origins of time asymmetry. Most important of all, this path to quantum theory removes the main obstacles to a much more classical view of quantum mechanics than is usually thought to be possible. It seems to solve the problem of nonlocality, for example, and to open the door to the kind of interpretation of quantum theory that Einstein always favored: a view in which there is still an objective world out there, and no mysterious role for observers.

This is a very dramatic claim, and readers are right to be skeptical. If there were a solution of this kind in quantum theory, after all, how could it have gone unnoticed for so long? The answer, I think, is this: the presuppositions this suggestion challenges are so deeply embedded in our ordinary ways of thinking that normally we simply don't notice them. If we do notice them, they seem so secure that the thought of giving them up seems crazy, even in comparison to the bizarre alternatives offered by quantum theory. Only by approaching these presuppositions from an angle which has nothing to do with quantum theory—in particular, by thinking about how they square with what we have discovered about the physical origins of time asymmetry—do we find that there are independent reasons to give them up. Suddenly, this way of thinking about quantum theory looks not just sane, but a natural consequence of other considerations.

What are these presuppositions? They involve notions such as causation and physical dependence. As we ordinarily use them, these notions are strongly time-asymmetric. For example, we take it for granted that events depend on earlier events in a way in which they do not depend on later events. Physicists often dismiss this asymmetry as subjective, terminological, or merely "metaphysical." As we shall see, however, it continues to exert a very powerful influence on their intuition—on what kind of models of the world they regard as intuitively acceptable. It is the main reason why the approach to quantum theory I want to recommend has received almost no serious attention.

In chapters 5–7 I mount a two-pronged attack on this intuition. Chapter 5 shows that it sits very uneasily with the kind of picture of the nature and origins of time asymmetry in physics which emerges from the earlier chapters. In this chapter I also explain in an introductory way why abandoning this intuition would have important and very attractive ramifications in the debate about quantum theory. However, the notions of causation, dependence, and the like are not straightforward. They are notions which have often puzzled philosophers, and their temporal asymmetry is especially mysterious. Is it some extra ingredient of the world, over and above the various asymmetries in physics, for example? Or can it be reduced to those asymmetries? These are philosophical issues, and the second arm of my attack on the intuition mentioned above involves an investigation of its origins, along philosophical lines.

In chapter 6 I argue that the asymmetry of causation cannot be reduced to any of the available physical asymmetries, such as the second law of thermodynamics. The basic problem for such a reduction is that the available physical asymmetries are essentially macroscopic, and therefore cannot account for causal asymmetry in microphysics—though our causal intuitions are no less robust when applied to this domain than they are elsewhere. I argue instead that the asymmetry of causation is anthropocentric in origin. Roughly, it reflects the time-asymmetric perspective we occupy as *agents* in the world—the fact that we deliberate for the *future* on the basis of information about the *past,* for example.

As I explain in chapter 7, this account has the satisfying consequence that despite its powerful grip on our intuitions—a grip which ought to seem rather puzzling, in view of the apparent symmetry of physics itself—causal asymmetry does not reflect a further ingredient of the world, over and above what is already described by physics. It doesn't multiply the objective temporal "arrows," in other words. More surprisingly, we shall see that the account does leave room for a limited violation of the usual causal order. In other words, it leaves open the possibility that the world might be such that from our standard asymmetric perspective, it would be appropriate to say that certain of our present actions could be the causes of *earlier* effects. In failing to recognize this possibility, physics has failed to practice what it has often preached concerning the status of causal asymmetry. Having often concluded, rightly, that the asymmetry of causation is not a physical matter, physicists have then failed to notice that the anthropocentric framework continues to constrain their construction of models of reality. One of the great attractions of the Archimedean standpoint is that it serves to break these conventional bonds, and hence to free physics from such self-imposed constraints.

The last two chapters apply these lessons to the puzzles of quantum mechanics. Chapter 8 provides an informal overview of the long debate about how quantum mechanics should be interpreted, identifying the main positions and their advantages and disadvantages. As I'll explain, the best focus for such an overview is the question that Einstein took to be the crucial one about quantum mechanics: Does it give us a complete description of the systems to which it applies?

Famously, Einstein thought that quantum theory is incomplete, and that there must be some further, more classical reality "in the background." His great disagreement with Niels Bohr centered on this issue. Einstein is often said to have lost the argument, at least in hindsight. (The work of John Bell in the 1960s is often thought to have put the final nail in Bohr's case, so to speak.) I think this verdict is mistaken. Despite Bell's work, Einstein's view is very much less implausible than it is widely taken to be, at least in comparison to the opposing orthodoxy.

This conclusion is overshadowed by that of chapter 9, however, where I show how dramatically the picture is altered if we admit the kind of backward causation identified in chapter 7. In the quantum mechanical literature this possibility is usually dismissed, or simply overlooked, because it flies in the face of such powerful intuitions about causality. But the lesson of chapter 7 is that when we ask where these intuitions come from, we discover that their foundations give us no reason at all to exclude the kind of limited backward influence in question—on the contrary, if anything, because powerful symmetry principles can be made to work in favor of the proposal.

In effect, then, my conclusion in chapter 9 is that the most promising and well-motivated approach to the peculiar puzzles of quantum mechanics has been almost entirely neglected, in part because the nature and significance of our causal intuitions have not been properly understood. Had these things been understood in advance—and had the real lessons of the nineteenth-century debate about temporal asymmetry been appreciated a century ago—then quantum mechanics is the kind of theory of microphysics that the twentieth century might well have expected.

REMARKS ON STYLE

A few remarks on the style and level of the book. Much of the argument is philosophical in character. It deals with live issues in contemporary physics, however, and takes for granted that it is physicists who need to be convinced of the advantages of the Archimedean standpoint. The book thus faces the usual hurdles of an interdisciplinary work, with the additional handicap of a far-reaching and counterintuitive conclusion. There is a danger that specialist readers on both sides will feel that my treatment of their own material is simplistic or simply wrong, and that my account of the other side's contribution is difficult, obscure and of doubtful relevance. Physicists are more likely to have the first reaction, of course, and philosophers the second, because I am writing from a philosophical standpoint.

There are conflicting constraints here, but the best approach seems to be to try to maximize clarity and readability, even if sometimes at the expense of rigor and precision. I have tried in particular to keep philosophical complexity to a minimum, in order to make the general viewpoint as accessible as possible to readers from other fields. On the physical side I had less choice in the matter—my own technical abilities soon reach their limits—but here too, where possible, I have tried to opt for accessibility rather than precision. Occasionally, where technicality of one sort or the other seemed especially important, I have tried to quarantine it, so that the details may be skipped by readers who are disinclined to tangle. (In these cases I indicate in the text which sections can be skipped.) Most chapters finish with a summary, and there is an overview of the book as a whole at the end.

Finally, a hint for impatient readers, keen to get into the quantum mechanics: start at chapter 5, and follow the arrows from there.

THE STOCK PHILOSOPHICAL DEBATES ABOUT TIME

The philosophy of time has a long history, and is unusual even by philosophical standards for the durability of some of its main concerns. In a modern translation much of Saint Augustine's work on time would pass for twentieth-century philosophy. Augustine's concerns are often exactly those of modern philosophers. He is puzzled about the nature of the distinctions between the past, the present, and the future, and about the fact that the past and the future seem unreal: the past has ceased to exist, and the future doesn't yet exist. And he is concerned about the nature and status of the apparent flow of time.

These two problems—the first the status of the past-present-future distinction, and the related concern about the existence of the past and the future, and the second the issue of the flow of time—remain the focus of much work in the philosophy of time. As I noted earlier, philosophers tend to divide into two camps. On one side there are those who regard the passage of time as an objective feature of reality, and interpret the present moment as the marker or leading edge of this advance. Some members of this camp give the present ontological priority, as well, sharing Augustine's view that the past and the future are unreal. Others take the view that the past is real in a way that the future is not, so that the present consists in something like the coming into being of determinate reality.

Philosophers in the opposing camp regard the present as a subjective notion, often claiming that *now* is dependent on one's viewpoint in much the same way that *here* is. Just as "here" means roughly "this place," so "now" means roughly "this time," and in either case what is picked out depends where the speaker stands. On this view there is no more an objective division of the world into the past, the present, and the future than there is an objective division of a region of space into here and there. Not surprisingly, then, supporters of this view deny that there is any ontological difference—any difference concerning simply *existence*—between the past, the present, and the future.

Often this is called the *block universe view,* the point being that it regards reality as a single entity of which time is an ingredient, rather than as a changeable entity set *in* time. The block metaphor sometimes leads to confusion, however. In an attempt to highlight the contrast with the dynamic character of the "moving present" view of time, people sometimes say that the block universe is *static*. This is rather misleading, however, as it suggests that there is a time frame in which the four-dimensional block universe stays the same. There isn't, of course. Time is supposed to be included in the block, so it is just as wrong to call it static as it is to call it dynamic or changeable. It isn't any of these things, because it isn't the right sort of entity—it isn't an entity *in* time, in other words.

Defenders of the block universe view deny that there is an objective present, and usually also deny that there is any objective flow of time. Indeed, perhaps the strongest reason for denying the objectivity of the present is that it is so difficult to make sense of the notion of an objective flow or passage of time. Why? Well, the stock objection is that if it made sense to say that time flows then it would make sense to ask how fast it flows, which doesn't seem to be a sensible question. Some people reply that time flows at one second per second, but even if we could live with the lack of other possibilities, this answer misses the more basic aspect of the objection. A rate of seconds per second is not a rate at all in physical terms. It is a dimensionless quantity, rather than a rate of any sort. (We might just as well say that the ratio of the circumference of a circle to its diameter flows at π seconds per second!)

A rarer but even more forceful objection is the following. If time flowed, then—as with any flow—it would only make sense to assign that flow a *direction* with respect to a choice as to what is to count as the positive direction of time. In saying that the sun moves from east to west or that the hands of a clock move clockwise, we take for granted the usual convention that the positive time axis lies toward what we call the future. But in the absence of some objective grounding for this convention, there isn't an objective fact as to which way the sun or the hands of the clock are "really" moving. Of course, proponents of the view that there is an objective flow of time might see it as an advantage of their view that it does provide an objective basis for the usual choice of temporal coordinate. The problem is that until we have such an objective basis we don't have an objective sense in which time is flowing one way rather than the other. In other words, not only does it not seem to make sense to speak of an objective *rate* of flow of time; it also doesn't make sense to speak of an objective *direction* of flow of time.

These problems in making sense of an objective flow of time spill over on the attempt to make sense of an objective present. For example, if the present is said to be the "edge" at which reality becomes concrete, at which the indeterminacy of the future gives way to the determinacy of the past, then the argument just given suggests that there isn't an objective sense in which reality is growing rather than shrinking.

These objections are all of a philosophical character, not especially dependent on physics. A new objection to the view that there is an objective present arises from Einstein's theory of special relativity. The objection is most forceful if we follow Augustine in accepting that only the present moment is real. For then if we want to inquire what reality includes, apart from our immediate surroundings, we need to think about what is *now* happening elsewhere. However, Einstein's theory tells us that there is no such thing as objective simultaneity between spatially separated events. Apparent simultaneity differs from observer to observer, depending on their state of motion, and there is no such thing as an objectively right answer. So the combination of Augustine and Einstein seems to give us the view that reality too is a perspective-dependent matter. The distinctive feature of the Augustinian view—the claim that the content of the present moment is an objective feature of the world—seems to have been lost.

Augustine's own reasons for believing in the objectivity of the present— indeed, the nonreality of everything else—seem to have been at least partly linguistic. That is, he was moved by the fact that we say such things as "There are no dinosaurs—they no longer exist" and "There is no cure for the common cold—it doesn't yet exist." By extrapolation, it seems equally appropriate to say that there is no past, for it no longer exists; and that there is no future, for it does not yet exist. However, a defender of the block universe view will say that in according these intuitions the significance he gives them, Augustine is misled by the tense structure of ordinary language. In effect, he fails to notice that "Dinosaurs do not exist" means "Dinosaurs do not exist *now.*" As a result, he fails to see that the basic notion of existence or reality is not the one that dinosaurs are here being said to lack—viz., existence *now*— but what we might term existence *somewhen.* Again the spatial analogy seems helpful: we can talk about existence in a spatially localized way, saying, for example, that icebergs don't exist here in Sydney; but in this case it is clear that the basic notion of existence is the unqualified one—the one that we would describe as existence *somewhere,* if language required us to put in a spatial qualification. We are misled in the temporal case because the simplest grammatical form actually includes a temporal qualification.

So it is doubtful whether Augustine's view can be defended on linguistic grounds. In practice, the most influential argument in favor of the objective present and objective flow of time rests on an appeal to psychology—to our own experience of time. It seems to us as if time flows, the argument runs, and surely the most reasonable explanation of this is that there is some genuine movement of time which we experience, or in which we partake.

Arguments of this kind need to be treated with caution, however. After all, how would things seem if it time didn't flow? If we suppose for the moment that there is an objective flow of time, we seem to be able to imagine a world which would be just like ours, except that it would be a four-dimensional block universe rather then a three-dimensional dynamic one. It is easy to see how to map events-at-times in the dynamic universe onto events-at-temporal-locations in the block universe. Among other things, our individual mental states get mapped over, moment by moment. But then surely our copies in the block universe would have the same experiences that we do—in which case they are not distinctive of a dynamic universe after all. Things would seem this way, even if we ourselves were elements of a block universe.

Proponents of the block universe view thus argue that in the case of the apparent flow of time, like that of the apparent objectivity of the present, it is important to draw a distinction between how things *seem* and how they actually are. Roughly speaking, what we need to do is to explain why things *seem* this way, without assuming that the "seeming" corresponds directly to anything in reality. Explanations of this kind are quite common in philosophy. Their general strategy is to try to identify some characteristic of the standpoint from which we "see" the appearance in question, such that the nature of the appearance can be explained in terms of this characteristic of the viewpoint. (There are lots of commonplace examples of this kind of thing. Rose-tinted spectacles explain why the world seems warm and friendly to those who wear them.)[4]

One of my projects in this book is to try to extend these insights about the consequences of the temporal perspective from which we view the world. We are interested in this partly for its bearing on the attempt to explain the arrow of time—existing attempts often go wrong because they fail to notice the influence of this perspective on ordinary ways of thinking—but also for its general philosophical interest. In this respect, as I said earlier, the book is an attempt to further the project of philosophical writers such as Williams, Smart, and Mellor.

From now on I shall simply take for granted the main tenets of the block universe view. In particular, I'll assume that the present has no special objective status, instead being perspectival in the way that the notion of *here* is. And I'll take it for granted that there is no objective flow of time. These assumptions will operate mainly in a negative way. I shall not explore the suggestion that flow gives direction to time, for example, because I shall be taking for granted that there is no such thing as flow.

In making these assumptions I don't mean to imply that I take the arguments for the block universe view sketched above to be conclusive. I do think that it is a very powerful case, by philosophical standards. However, the aim of the book is to explore the consequences of the block universe view in physics and philosophy, not to conduct its definitive defense. My impression is that these consequences give us new reasons to favor the view over its Augustinian rival, but others might take the point in reverse, finding here new grounds for the claim that the block universe leaves out something essential about time. Either way, all that matters to begin with is that the block universe view is not already so implausible that it would a waste of time to seek to extend it in this way, and this at least is not in doubt.

THE ARROWS OF TIME

Our main concern is with the asymmetry of time, but what does this mean? The terminology suggests that the issue concerns the asymmetry *of time itself,* but this turns out not to be so. To start with, then, we need to distinguish the issue of the asymmetry *of* time from that of the asymmetry of things *in* time. The easiest way to do this is to use a simple spatial analogy.

Imagine a long narrow table, set for a meal. The contents of the table might vary from end to end. There might be nonvegetarian food at one end and vegetarian at the other, for example; there might be steak knives at one end but not at the other; all the forks might be arranged so as to point to the same end of the table; and so on. This would constitute asymmetry *on* the table. Alternatively, or as well, the table itself might vary from end to end. It might be wider or thicker at one end than the other, for example, or even bounded in one direction but infinite in the other. (This might be a meal on Judgment Day, for example, with limited seating at the nonvegetarian end.) These things would be asymmetries *of* the table—asymmetries of the table itself, rather than its contents.

There seems to be an analogous distinction in the case of time. Time itself might be asymmetric in various ways. Most obviously, it might be bounded in one direction but not in the other. There might be an earliest time but no latest time. There are other possibilities: as long as we think of time as a kind of extended "stuff," there will be various ways in which the characteristics of this stuff might vary from end to end. More contentiously, if sense could be made of the notion of the flow of time, then that too might provide a sense in which time itself had an intrinsic direction or asymmetry. (However, supporters of the objective present/objective flow view are likely to be unhappy with this use of a spatial metaphor to characterize the distinction between the asymmetry of time and that of things in time.)

Independently of the issue as to whether time itself is symmetric from end to end, there is an issue about whether the physical contents of time are symmetric along its axis. This is analogous to the question as to whether the contents of the table are symmetric from end to end. It turns out that the interesting questions about temporal asymmetry are very largely of this kind.

There are various respects in which the contents of the block universe appear to be arranged asymmetrically with respect to the temporal axis. For example, many common physical processes seem to exhibit a very marked temporal preference, occurring in one temporal orientation but not the other. This is why the events depicted in reversed films often seem bizarre. In the real world, buildings may collapse into rubble, for example, but rubble does not "uncollapse" to form a building—even though, as it happens, the latter process is no less consistent than the former with the laws of mechanics. (It is this last fact that makes the asymmetry so puzzling—more on this in a moment.)

As we shall see in the following chapters, there are a number of apparently distinct ways in which the world we inhabit seems asymmetric in time. One of the tasks of an account of temporal asymmetry is thus a kind of taxonomic one: that of cataloging the different asymmetries (or "arrows," as they have come to be called), and sorting out their family relationships. Physicists in particular have been interested in the question as to whether there is a single "master arrow," from which all the others are in some sense derived. As we shall see, the leading candidate for this position has been the so-called arrow of thermodynamics. This is the asymmetry embodied in the second law of thermodynamics, which says roughly that the entropy of an isolated physical system never decreases.

As a gentle introduction to the kind of reasoning on which much of the book depends, note that this formulation of the second law assumes a choice of temporal orientation. It assumes that we are taking the "positive" temporal direction to be that of what we ordinarily call the future. There is nothing to stop us taking the positive axis to lie in the opposite direction, however, in which case the second law would need to be stated as the principle that the entropy of an isolated system never *increases*. The lesson is that the objective asymmetry consists in the presence of a unidirectional gradient in the entropy curve of, apparently, all isolated physical systems. Each such system exhibits such a gradient, and all the gradients slope in the same temporal direction. But it is not an objective matter whether the gradients *really* go up or go down, for this simply depends on an arbitrary choice of temporal orientation. They don't *really* go either way, from an atemporal viewpoint.

THE PUZZLE OF ORIGINS

One of the problems of temporal asymmetry is thus to characterize the various temporal arrows—asymmetries of things *in* time—and to explain how they relate to one another. Let's call this the *taxonomy problem*. The second problem—call it the *genealogy problem*—is to explain why there is *any* significant asymmetry of things in time, given that the fundamental laws of physics appear to be (almost) symmetric with respect to time. Roughly, this symmetry amounts to the principle that if a given physical process is permitted by physical laws, so too is the reverse process—what we would see if a film of the original process were shown in reverse. With one tiny exception—more on this in a moment—modern physical theories appear to respect this principle. This means that insofar as our taxonomy of temporal arrows reveals significant asymmetries—significant cases in which the world shows a preference for one temporal orientation of a physical process over the other, for example—it is puzzling how these asymmetries could be explained in terms of the available physical theories. How are we going to explain why buildings collapse into rubble but rubble does not "uncollapse" into buildings, for example, if both processes are equally consistent with the laws of mechanics? We seem to be trying to pull a square rabbit from a round hat!

As I noted, however, there seems to be one little exception to the principle that the basic laws of physics are time-symmetric. This exception, first discovered in 1964, concerns the behavior of a particle called the neutral kaon. To a very tiny extent, the behavior of the neutral kaon appears to distinguish past and future—an effect which remains deeply mysterious.[5] Tiny though it is, could this effect perhaps have something to do with the familiar large-scale asymmetries (such as the tendency of buildings to collapse but not "uncollapse")? At present, it is difficult to offer a convincing answer to this question, one way or the other. The best strategy is to set the case of the kaon to one side, and to study the more familiar arrows of time in physics as if there were no exceptions to the principle that the underlying laws are time-symmetric. This way we can find out where the puzzles really lie—and where, if at all, the kaon might have a role to play.[6]

Physicists and philosophers have long been puzzled by the genealogy problem. The most famous attempt to provide at least a partial solution dates from the second half of the nineteenth century, when Boltzmann claimed to have derived the second law of thermodynamics for the case of gases from a statistical treatment within the symmetrical framework of Newtonian mechanics. As we shall see in the next chapter, however, Boltzmann's critics soon pointed out that he had relied on a temporally asymmetric assumption (the so-called *stoßzahlansatz*, or "assumption of molecular chaos"). Boltzmann's argument thus provides an early example of what has proved a common and beguiling fallacy. In search of an explanation for the observed temporal asymmetries—for the observed difference between the past and the future, in effect—people unwittingly apply different standards with respect to the two temporal directions. The result is that the asymmetry they get out is just the asymmetry they put in. Far from being solved, the problems of temporal asymmetry are obscured and deferred—the lump in the carpet is simply shifted from one place to another. In the course of the book we shall encounter several examples of this kind of mistake.

The reason the mistake is so prevalent is not (of course) that the physicists and philosophers who have thought about these problems are victims of some peculiar intellectual deficit. It is simply that temporal asymmetry is so deeply ingrained in our ways of thinking about the world that it is very difficult indeed to spot these asymmetric presuppositions. Yet this is what we need to do, if we are to disentangle the various threads in the problem of temporal asymmetry, and in particular to distinguish those threads that genuinely lie in the world from those that merely reflect our own viewpoint. In order to explain temporal asymmetry it is necessary to shake off its constraints on our ordinary ways of thinking—to stand in thought at a point outside of time, and thence to regard the world in atemporal terms. This book is a kind of self-help manual for those who would make this Archimedean journey.

To put the project in perspective, let us reflect again on the history of science, or natural philosophy more generally. In hindsight it is easy to see that our view of the world has often unwittingly embodied the peculiarities of our own standpoint. As I noted earlier, some of the most dramatic episodes in the history of science are associated with the unmasking of distortions of this kind. I mentioned Copernicus and Darwin. Another striking example is the conceptual advance that led to Newton's first law of motion. This advance was Galileo's appreciation that the friction-dominated world of ordinary mechanical experience was not the natural and universal condition it had been taken to be. Left to its own devices, a moving body would move forever.

In the same historical period we find a parallel concern with the philosophical aspects of the project of uncovering the anthropocentricities of our ordinary view of the world. We find an interest in what soon came to be called the distinction between primary and secondary qualities, and an appreciation that the proper concern of physics is with the former: that is, with those aspects of the world that are not the product of our own perceptual peculiarities.

Consider these remarks from Galileo himself, for example, in 1623:

I feel myself impelled by the necessity, as soon as I conceive a piece of matter or corporeal substance, of conceiving that in its own nature it is bounded and figured in such and such a figure, that in relation to others it is large or small, that it is in this or that place, in this or that time, that it is in motion or remains at rest, that it touches or does not touch another body, that it is single, few, or many; in short by no imagination can a body be separated from such conditions; but that it must be white or red, bitter or sweet, sounding or mute, of a pleasant or unpleasant odour, I do not perceive my mind forced to acknowledge it necessarily accompanied by such conditions; so if the senses were not the escorts, perhaps the reason or the imagination by itself would never have arrived at them. Hence I think that these tastes, odours, colours, etc., on the side of the object in which they seem to exist, are nothing else than mere names, but hold their residence solely in the sensitive body; so that if the animal were removed, every such quality would be abolished and annihilated.[7]

Galileo is telling us that tastes, odors, colors, and the like are not part of the objective furniture of the world; normally, in thinking otherwise, we mistake a by-product of our viewpoint for an intrinsic feature of reality. In Galileo and later seventeenth-century writers, the move to identify and quarantine these secondary qualities is driven in part by the demands of physics; by the picture supplied by physics of what is objective in the world. This is not a fixed constraint, however. It changes as physics changes, and some of these changes themselves involve the recognition that some ingredient of the previously excepted physical world view is anthropocentric.

These examples suggest that anthropocentrism infects science by at least two different routes. In some cases the significant factor is that we happen to live in an exceptional part of the universe. We thus take as normal what is really a regional specialty: geocentric gravitational force, or friction, for example. In other cases the source is not so much in our *location* as in our *constitution*. We unwittingly project onto the world some of the idiosyncrasies of our own makeup, seeing the world in the colors of the in-built glass through which we view it. But the distinction between these sources is not always a sharp one, because our constitution is adapted to the peculiarities of our region.

It is natural to wonder whether modern physics is free of such distortions. Physicists would be happy to acknowledge that physics might uncover new locational cases. Large as it is, the known universe might turn out to be an unusual bit of something bigger.[8] The possibility of continuing constitutional distortions is rather harder to swallow, however. After all, it challenges the image physics holds of itself as an objective enterprise, an enterprise concerned with not with how things *seem* but with how they actually *are*. It is always painful for an academic enterprise to have to acknowledge that it might not have been living up to its own professed standards!

In the course of the book, however, I want to argue that in its treatment of time asymmetry, contemporary physics has failed to take account of distortions of just this constitutional sort—distortions which originate in the kind of entities we humans are, in one of our most fundamental aspects. If we see the historical process of detection and elimination of anthropocentrism as one of the adoption of progressively more detached standpoints for science, my claim is that physics has yet to achieve the standpoint required for an understanding of temporal asymmetry. In this case the required standpoint is an atemporal one, a point outside time, a point free of the distortions which stem from the fact that we are creatures in time—truly, then, a view from nowhen.

2

$\longleftarrow\longrightarrow$

"More Apt to Be Lost than Got":
The Lessons of the Second Law

In the last chapter I distinguished two kinds of temporal asymmetry: the asymmetry of time itself, on the one hand, and the asymmetry of things *in* time, on the other. I compared this to the distinction between the asymmetry of an elongated table from end to end, and the asymmetry of the contents of the table along the same axis. I noted that in practice the latter kind of asymmetry turns out to be the interesting one. The striking temporal arrows of the real world all seem to be of this kind.

I also distinguished two kinds of issue about these real-world asymmetries: the taxonomic issue as to how many arrows there are, and how they relate to one another, and the genealogical issue as to how such dramatic asymmetries come to exist in a world whose underlying physical laws seem to be time-symmetric. In this and the two following chapters we shall be looking at what modern physics has to say about these issues.

This chapter is concerned with the most famous temporal arrow in physics, that of the second law of thermodynamics—"the universal tendency of entropy to increase," as one early formulation puts it. We shall see how the tension between the temporal asymmetry of thermodynamics and the symmetry of Newtonian mechanics came gradually to the attention of physics in the second half of the nineteenth century. In describing some of the various attempts that were made to reconcile these theories, I shall focus in particular on the tendency of these attempts to make a very characteristic mistake: that of assuming temporal asymmetry in some disguised or unrecognized form. I want to argue that this mistake is itself a product of a failure to adopt an atemporal perspective on the problem. Mistakes which are almost invisible from the ordinary temporal viewpoint are thrown into stark relief once we familiarize ourselves with the Archimedean standpoint.

The nineteenth-century debate wasn't entirely confused about these points, of course. As we'll see, the great Austrian physicist Ludwig Boltzmann (1844–1906) made a major advance toward an atemporal view of the puzzles of temporal asymmetry. But even in Boltzmann's case the advance was incomplete, and we'll see that from the 1890s to the present, physicists and philosophers have usually failed to grasp the true moral of the issues raised at that time. In particular, the debate has remained in the grip of a misleading conception of the true nature of the problem—a mistaken view of what actually needs to be explained about the asymmetry of thermodynamics. As a result, much contemporary work on these issues is still misdirected.

Above all, then, the goals of this chapter are to provide a clear account of what it is that actually needs to be explained about the temporal asymmetry of thermodynamics, and to offer the Archimedean strategy as a remedy for the kind of mistakes that have led so many physicists and philosophers on wild goose chases across this territory in the past 120 years.

IRREVERSIBILITY DISCOVERED: NEWTON TO BOLTZMANN

Science is very selective about what it tries to explain. It often happens that commonplace phenomena are simply not noticed by science until, in virtue of some new hypothesis or theoretical advance, they suddenly appear to be problematic and in need of explanation. (Perhaps more surprisingly, the opposite thing can also happen. Science can decide that a class of phenomena that appeared to require explanation are not problematic after all. More on this below.) Something of this kind seems to have taken place in the eighteenth and early nineteenth centuries, with respect to a wide class of phenomena that came to be seen as instances of a general principle termed the second law of thermodynamics.

It might seem that the fact that many physical processes are irreversible is so obvious that it is ridiculous to speak of its being discovered by science. As the science historian Stephen Brush puts it, "It is difficult to conceive of a time when people did not know that heat flows from hot bodies to cold bodies."[1] The notion of irreversibility requires a contrast, however, so that in another sense science was not equipped to notice *irreversibility* until it noticed *reversibility,* or at least temporal *constancy.*

Developments in the seventeenth century thus become crucial. Isaac Newton himself was interested in processes involving the irreversible dissipation of motion, particularly in connection with the issue as to whether a mechanistic universe requires continual Divine intervention. His view was that it does, for in the Heavens, as in general, "Motion is much more apt to be lost than got, and is always upon the decay."[2] However, while it is the reversible character of Newtonian mechanics that becomes crucial later in our story, the emergence of the tension between this reversibility on the one hand, and the irreversibility of many ordinary phenomena on the other, had to await a more direct connection between the mechanics and the phenomena concerned.[3]

By the end of the eighteenth century, however, science was very familiar with a range of different kinds of phenomena with the common property that they did not seem to occur in reverse: the diffusion of gases, the dissipative effects of frictional forces, and the flow of heat from warm bodies to cold, for example. All the same, the idea that these processes required explanation in terms of some general principle or natural law seems to have been slow in coming. Perhaps part of the explanation for this is that irreversibility is *so* familiar that it is difficult to see it as something that requires explanation. Brush records scattered statements of general principles from the late eighteenth and early nineteenth centuries, the earliest being that of John Hunter in 1788: "It is well known that heat in all bodies has a tendency to diffuse itself equally through every part of them, till they become of the same temperature."[4]

The development of thermodynamics in the early nineteenth century provided a unifying theoretical framework within which many if not all of the familiar irreversible processes could eventually be seen to exemplify the same general tendency—"a universal tendency in nature to the dissipation of mechanical energy," as William Thomson (1824–1907) describes it in a note of that title in 1852,[5] or "the universal tendency of entropy to increase," as Rudolf Clausius (1822–1888) puts it in the first modern characterization of the second law of thermodynamics, in 1865.[6] The theory brought with it the crucial notion of *equilibrium*—the state toward which a system tends, when it is left to its own devices. Note that this in itself is an asymmetric notion, at least at first sight. It concerns what is liable to happen to a system in the future, not what happens to it in the past. However, we shall see that this asymmetry came to be called into question, in the debates that followed the clash between the irreversibility of thermodynamics and the reversibility of Newtonian mechanics.

The new theory of thermodynamics enjoyed a brief period of calm around midcentury, during which its conflict with Newtonian mechanics was not yet manifest. At this time it was still a defensible view that thermodynamics described an autonomous range of phenomena, not in need of reduction to mechanics. This avenue was soon blocked, however, as Newtonian methods came to be applied to the paradigm phenomena of thermodynamics. The decisive advances were the development of the kinetic theory of heat—the theory that characterizes heat in terms of molecular motion—and its extension to explain other features of the behavior of gases in statistical terms.

With the success of these theories it was no longer reasonable to deny that the phenomena described by thermodynamics were a manifestation of the mechanical behavior of the microscopic constituents of matter.

The crucial link between the statistical ideas of the kinetic theory of gases and the concepts of thermodynamics was forged by the brilliant and prolific Scottish physicist James Clerk Maxwell (1831–1879). In his 1867 paper "On the Dynamical Theory of Gases," Maxwell shows how to derive a number of characteristics of the macroscopic behavior of gases from a theory concerning the statistical distribution of the velocities of their component molecules. Most important for our purposes, he derives an expression for the distribution of velocities for a gas in thermal equilibrium. The case for regarding the distribution concerned as that which corresponds to thermal equilibrium is indirect: Maxwell argues that a gas that has the distribution of velocities concerned will not depart from it, and offers an informal argument that no other distribution has this property. Since thermal equilibrium is by definition a stable state, it follows that it is characterized by the Maxwell distribution.

Maxwell's argument for the stability of the Maxwell distribution relies on an assumption whose significance only becomes apparent later, when the issue of irreversibility comes more clearly into focus. He assumes that the number of collisions between molecules of given velocities will simply be proportional to the product of the number having one velocity and the number having the other. Thus if 1 in every 1,000 molecules has velocity v_1 and 1 in 1,000 has velocity v_2, the assumption implies that 1 in 1,000,000 collisions involves molecules with velocities v_1 and v_2; the velocities of colliding molecules are independent, in other words. This assumption may seem innocuous, but it comes to play a crucial role, as we shall see.

The next step belongs to Boltzmann, in whose work the connection between statistical mechanics and the second law of thermodynamics first becomes absolutely explicit. Like Maxwell, Boltzmann considered the effects of collisions on the distribution of velocities of the molecules of a gas. In effect, his approach was to partition the available velocities into a series of tiny intervals, and to consider the effect of collisions on the number of molecules whose velocities fell in each of these intervals. He was able to argue that no matter what the initial distribution of velocities—that is, no matter how the particles were initially assigned to the available velocity intervals—the effect of collisions was to make the distribution of velocities approach the distribution which Maxwell had characterized a few years earlier.

Boltzmann's approach was to define a quantity, originally denoted E, with respect to which he was able to argue (1) that E takes its minimum possible value when the gas has the Maxwell distribution, and (2) that when the gas does not already have the Maxwell distribution, the effect of collisions is to ensure that E always decreases with time. Together these results seemed to imply not only that a gas will not depart from the Maxwell distribution, but also that any sample of gas will tend to that distribution, if not already there. These are the defining characteristics of the notion of equilibrium, however, and Boltzmann's result therefore provides powerful confirmation that the Maxwell distribution does indeed characterize the condition of a gas in thermal equilibrium, as Maxwell himself had suggested. (E later came to be called *H,* and the result is now known as Boltzmann's *H*-theorem.)

Moreover, the equilibrium-seeking behavior of gases was an instance of the general irreversible tendency described by Clausius's second law, and Boltzmann saw that there was a direct connection between the *H*-theorem and Clausius's principle. In the case of a gas already in equilibrium, Boltzmann's quantity *H* was equivalent to minus the entropy, as defined by Clausius. Boltzmann therefore suggested that *H* provided a generalized notion of entropy. In showing that *H* always decreases, the *H*-theorem thus amounted to a proof of a generalized second law for the case of gases; in other words, it amounted to a proof that entropy always *increases.*

The mechanics underlying Boltzmann's *H*-theorem are time-symmetric in themselves, however. How then does he manage to derive a time-asymmetric conclusion, namely that *H decreases* over time? Not surprisingly, the answer is that he has actually imported a time-asymmetric assumption. It is his *stoßzahlansatz,* or "assumption of molecular chaos." In effect, this is Maxwell's assumption that the probabilities (or, more strictly, frequencies) of velocities of colliding particles are independent. A pair of particles is not more or less likely to have a particular combination of initial velocities in virtue of their collision than they would be otherwise: the frequency of collisions between particles with velocities v_1 and v_2 is just the product of the relative frequency with which particles have velocities v_1 and v_2.

Why is this an asymmetric assumption? Because we expect the velocities of particles to become correlated *as a result* of their collisions. We expect outgoing products of collisions to be correlated with one another in various ways, *even if they never encounter one another in the future.* We do not expect the incoming components of a collision to be correlated, *if they have never encountered one another in the past.* Boltzmann's assumption thus appears to be an instance of a very plausible general principle, which I'll call the Principle of the Independence of Incoming Influences, or PI^3, for short. We'll encounter PI^3 at various points in this book, for it is one of the main forms in which asymmetry slips almost unnoticed into attempts to account for the arrows of time. For the moment what needs to be emphasized is that insofar as Boltzmann's general program employing the *H*-theorem is on the right track, it doesn't so much explain the temporal asymmetry embodied in the second law of thermodynamics, as shift the problem from one place to another. If the *stoßzahlansatz* weren't an asymmetric assumption, Boltzmann's theorem wouldn't yield an asymmetric conclusion. So the effect of the *H*-theorem is to give us the puzzle of the asymmetry of the *stoßzahlansatz,* in place of the puzzle of the asymmetry of the second law itself.

Shifting the problem in this way is not necessarily a useless exercise. On the contrary, in trying to track down the origins of temporal asymmetry one of our main tasks is likely to be that of chasing it from its less basic to its more basic manifestations, a process which quite properly has the effect of shifting the explanatory burden. What we need to guard against, however, is the tendency to overlook the problem in its new form.

The best defense against this tendency is a resolute commitment to the atemporal viewpoint. As we noted briefly in chapter 1, to adopt such a viewpoint in this case is to note that what needs to be explained is not that entropy *increases,* for the ordinary temporal perspective has no more objective validity than the reverse one, from which standpoint the question would be: Why does entropy always *decrease*? The objective issue is the one we obtain by taking the common factor of these alternatives, namely the question: Why is entropy uniformly higher in one direction than in the other? Once the question is explicitly posed in these terms, Boltzmann's answer comes to something like "Because we have molecular chaos in one direction but not the other"—and now it is obvious that we have simply shifted the lump in the carpet.

Historically, however, criticism of Boltzmann's claim to have explained the second law took a rather different path. A concern about the possibility of deriving an asymmetric conclusion from apparently symmetric theory did soon surface, but it did not initially focus on the assumption of molecular chaos, and was entangled, as we shall see, with a different critical point. By the time concerns about the derivation of asymmetry of the *H*-theorem did eventually turn to the role of molecular chaos, Boltzmann's statistical ideas had already taken a different tack. As we shall see, however, PI³ continued to play a crucial role in later attempts to explain the asymmetry of thermodynamics, and yet its own asymmetry continued—and continues—to be overlooked or ignored.

THE REVERSIBILITY OBJECTION I

When originally formulated in the mid-nineteenth century, the second law of thermodynamics was thought of as an exceptionless principle, on a par with the other laws of physics. As statistical approaches to thermodynamics developed, however, it came to be realized that the principles they implied would have a different character. Violations of statistical principles based on the average movement of vast numbers of particles might be highly unlikely, but they were not impossible. Far from excluding them, in fact, the statistical treatment underwrote their physical possibility.

Concerning the behavior of gases, this point was noticed by Maxwell in the late 1860s. Maxwell's Demon is an imaginary creature who segregates the fast and slow molecules of a gas, thereby making it hot in one region and cold in another. The point of the story is that what the Demon does might happen by accident, and that therefore the tendency of gases to reach thermal equilibrium can be nothing more than a tendency—it cannot be an exceptionless law. Even more tellingly, in discussions between Maxwell, Thomson, and P. G. Tait at this time, it was appreciated that for any possible gas in the process of approaching equilibrium, there is another in the process of departing from equilibrium: namely, the gas we get by exactly reversing the motions of all the molecules of the original. The determinism of Newtonian mechanics implies that this new state will simply retrace the path of the old state, so that if the gas had originally been becoming more uniform, the new state will lead to its becoming less uniform. Again, the conclusion is that if the second law is to be grounded on a statistical treatment of the behavior of the microscopic constituents of matter, it cannot be an exceptionless principle.

The notion of complete reversal of particle motions also occurred to Franz Loschmidt, a colleague of Boltzmann in Vienna, and it is by this route that the so-called reversibility paradox came to Boltzmann's attention. Contemporary writers sometimes note with approval an ironic response to this objection which tradition attributes to Boltzmann: "Go ahead and reverse them!" he is supposed to have said. Before we turn to Boltzmann's actual response, which is rather more subtle, it is worth noting that the atemporal viewpoint provides an irony-proof version of the objection. Given that there isn't an objective direction of time, we don't need to reverse anything to produce *actual* counterexamples of the imagined kind. Any actual case of entropy increase is equally a case of entropy decrease—which is to say that it isn't objectively a case of either kind, but simply one of entropy *difference* over time. (The argument doesn't really need actual examples, for it is only trying to establish the physical *possibility* of entropy-decreasing microstates. Lots of things are possible which never actually happen, like winning the lottery ten times in a row. However, to understand the atemporal viewpoint is to see that actual examples are all around us.)

In practice the argument took a different course, for Boltzmann's opponents do not appear to have questioned the objectivity of the direction of time (though he himself came to do so, as we shall see). The reversibility objection seems to have been seen by Maxwell, Thomson, and Loschmidt as an argument against the exceptionless character of the second law, not as an attack on Boltzmann's claim to have derived a time-asymmetric conclusion.

Boltzmann himself seems to have been at least partially aware of its significance in the latter respect, however. In responding to Loschmidt he adds the following note:

> I will mention here a peculiar consequence of Loschmidt's theorem, namely that when we follow the state of the world into the infinitely distant past, we are actually just as correct in taking it to be very probable that we would reach a state in which all temperature differences have disappeared, as we would be in following the state of the world into the distant future.[7]

Boltzmann's thought took a new direction in response to the reversibility objection. As we shall see, his later views embody a very subtle appreciation of the relationship between statistics and temporal asymmetry, and—at least at times—a grasp of the atemporal perspective which has rarely been matched in the succeeding century.

ENTROPY AS PROBABILITY

The reversibility objection convinced Boltzmann that his approach had the consequence that the second law is of a statistical rather than an exceptionless nature. He moved on from this to embrace a new conception of the nature of entropy itself. Applied to the case of gases, the crucial idea is that a given condition of the gas will normally be realizable in many different ways: for any given description of the gas in terms of its ordinary or "macroscopic" properties, there will be many different *microstates*—many different arrangements of the constituent molecules—which would produce the *macrostate* concerned.

Consider a sample of gas which is confined to two connected chambers, for example. One macroscopic property of the sample concerns the proportion of the gas which occupies each chamber. At a particular moment 52 percent of the gas might be in chamber A and 48 percent in chamber B, for example. There is a huge number of different arrangements of the individual molecules of the gas which would make this the case. These are the microstates which are said to "realize" the given macrostate. However, macrostates are far from uniform in this respect. Some can be realized in many more ways than others. If the chambers are of the same size and the gas is free to move between them, for example, there are comparatively few microstates which make the distribution of gas very uneven. This is exactly analogous to the fact that if we toss a coin 1,000 times, there are a lot more ways to get 500 heads than to get only 5 heads.

Hence if all possible microstates are assumed to be equally likely, the gas will spend far more of its time in some macrostates than in others. It will spend a lot of time in macrostates that can be realized in very many ways, and little time in macrostates that can be realized in very few ways. From here it is just a short step to the idea that the equilibrium states are those of the former kind, and that the entropy of a macrostate is effectively a measure of its probability, in these microstate-counting terms. Why then does entropy tend to increase, on this view? Simply because from a given starting point there are very many more microstates to choose from that correspond to higher entropy macrostates, than microstates that correspond to lower entropy macrostates.

This account builds the statistical considerations in from the start. Hence it makes explicit the first lesson of the reversibility objection, viz., that the second law is not exceptionless. Moreover, it seems to bypass the *H*-theorem—and hence in particular the assumption of molecular chaos—by attributing the general increase of entropy in gases not to the effects of collisions as such, but to broader probabilistic considerations. Where does the asymmetry come from, however, if not from the *stoßzahlansatz?*

The answer, as Boltzmann himself saw, is that there is *no* asymmetry in this new statistical argument. The above point about entropy increase toward (what we call) the future applies equally toward (what we call) the past. At a given starting point there are very many more possible histories for the gas that correspond to higher entropy macrostates in its past, than histories that correspond to lower entropy macrostates. Insofar as the argument gives us reason to expect entropy to be higher in the future, it also gives us reason to expect entropy to have been higher in the past. Suppose we find our gas sample unevenly distributed between its two chambers at a particular time, for example. If we consider the gas's possible future, there are many more microstates which correspond to a more even distribution than to a less even distribution. Exactly the same is true if we consider the gas's possible past, however, for the statistical argument simply relies on counting possible combinations, and doesn't know anything about the direction of time.

Boltzmann appreciated this point, and in a crucial advance also saw that the statistical treatment of entropy thus makes it extremely puzzling why entropy is *not* higher in the past. In fact, entropy seems to have been even lower in the past than it is now, in stark conflict with this statistical expectation. Why should this be so? At least to some degree, Boltzmann came to see that this puzzle represents the deep mystery of the subject. He suggested a novel solution, to which we'll turn in a moment.

THE REVERSIBILITY OBJECTION II

Thus Maxwell, Thomson, and Loschmidt all seem to have taken the point of the reversibility objection to be that the second law cannot be a universal principle, on a par with the laws of mechanics. Boltzmann himself saw its deeper significance—a challenge to *asymmetry* rather than the *universality* of the *H*-theorem and the second law—but the point seems to have failed to take root in the 1870s. In was revived with new vigor some ten years later, however, particularly by Edward Culverwell, of Trinity College, Dublin.[8] Culverwell seems to have seen more clearly than any of his predecessors the paradoxical nature of the claim to have derived an asymmetric conclusion—even of a statistical nature—from the symmetric foundations of classical mechanics. (His own suggestion was that an adequate mechanical foundation for thermodynamics would need to invoke an asymmetric effect of the aether on the particles of matter.)

The new debate focused renewed attention on the status of Boltzmann's *stoßzahlansatz*. Recall that this assumption amounts to what I called PI[3]—the principle that the properties of incoming colliding particles are independent of one another. For all its intuitive plausibility, this is a time-asymmetric principle. To rest a derivation of the second law on its shoulders is to leave the temporal asymmetry of thermodynamics no less mysterious than before.

Culverwell's intervention gave rise to a vigorous discussion, especially in the correspondence columns of *Nature* in 1894 and 1895. The debate makes interesting reading, not least because its key issues have remained unresolved and topical to this day. (The concerns of some of the other writers of the period have not fared so well; there is a series of letters on cannibalism in snakes, for example.) The discussion illustrates that it is remarkably difficult to keep one's eye on the ball in this game. Leaving aside commentators who miss the point completely, we find many astute thinkers being sidetracked. One of the most astute contributors is S. H. Burbury, who draws attention to the crucial role of the independence assumption.[9] Even in Burbury's hands, however, the problem is seen as that of justifying the claim that the independence assumption *continues to hold*, after the molecules of a given volume of gas have begun colliding with one another; the concern is that *previous* collisions will induce correlations between the participants in *future* collisions. This isn't a concern about the asymmetry reflected in PI[3] itself, however, for it doesn't raise any objection to the assumption that the molecules are uncorrelated before they begin to interact.

This point is borne out by Burbury's own positive suggestion, which is that molecular chaos is maintained in practice by the fact that "any actual material system receives disturbances from without, the effect of which, coming at haphazard, is to produce that very distribution of coordinates which is required to make H diminish."[10] The idea that the key to the general increase in entropy lies in the inevitable disturbance of systems by their environment has proved a popular one in the 100 years since Burbury first suggested it. Like Burbury, however, many more recent writers miss the crucial point, which is that the assumption that external influences "come at haphazard" is itself an instance of the general principle PI^3, which no one thinks of applying in reverse. In other words, we take it for granted that the influences in the other direction—the minute influences that a system inevitably exerts on its environment—do not "go at haphazard," but are rather correlated in a way which reflects their common origin. Yet if we view the same events from the reverse temporal perspective, it is these correlated influences which appear to be incoming, and Burbury's original disturbances which appear outgoing. Why should we assume that the latter now "go at haphazard"?

As I say, this point does not seem to have been made during the discussion in the 1890s, and has rarely been made since. It is interesting to note that it is missed even by Boltzmann, who at this period in particular shows considerable sensitivity to the issue of the objectivity of temporal direction. Boltzmann's own response to Burbury is to acknowledge the crucial role of the independence assumption, but to suggest that Burbury's concern about the correlating effect of prior collisions can be met by assuming that gas molecules usually travel a long way between collisions, compared to the average distance between neighboring molecules.[11] The idea is that this condition ensures that molecules will not normally be colliding with molecules which they have encountered in the recent past, and hence that the effects of correlations due to previous collisions will be insignificant. Again, however, this reply simply ignores the more basic temporal asymmetry of the independence assumption.

Ironically, however, it is far from clear that Boltzmann need be defending the H-theorem at this point. To see why this is so, we need to examine more closely what was involved in Boltzmann's recognition that there is a puzzle as to why entropy is low in the past.

BOLTZMANN'S SYMMETRIC VIEW

We have seen that in responding to Loschmidt's version of the reversibility objection in 1877, Boltzmann already understood that the statistical considerations he was introducing were symmetric in their implications. He notes the "peculiar consequence of Loschmidt's theorem," to the effect that on probabilistic grounds alone, we should expect entropy to increase toward the past, as well as toward the future. He must also have appreciated that to the extent to which we are entitled to assume that all microstates are equally probable—the foundational assumption of the probabilistic approach—it follows that the current low-entropy state of the world is itself highly improbable (and the lower entropy state we believe it to have had in the past even more so). Thus the statistical approach brings with it two rather unwelcome consequences. To the extent that it explains why entropy normally increases toward (what we call) the future, it also predicts that entropy should increase toward (what we call) the past; and it makes it rather mysterious why entropy should be so low *now.*

Boltzmann returned to these ideas in 1895 and 1896, prompted both by the debate initiated by Culverwell and by a new objection raised by the mathematicians Henri Poincaré and Ernst Zermelo. Poincaré had shown in 1889 that entropy-reducing behavior is not merely *possible,* as is established by the first reversibility objection, but *inevitable,* given enough time. Under certain general conditions, a physical system is bound eventually to come arbitrarily close to any possible state, even one of very low entropy. (This follows from Poincaré's so-called recurrence theorem.) In 1896 Zermelo argued that this result is incompatible with the attempt to account for the asymmetry of thermodynamics in mechanical terms.[12]

In effect, however, this recurrence argument merely confirmed the view to which Boltzmann had already been led in the 1870s, in response to the reversibility objection. In conceding that the second law is of a statistical rather than a universal nature, Boltzmann had already recognized that exceptions were not impossible, but merely improbable; and it is an elementary consequence of probability theory that even very improbable outcomes are very likely to happen, if we wait long enough. In effect, then, Boltzmann had already accepted the conclusion for which Zermelo now argued. Poincaré's recurrence theorem simply puts this implication of the statistical view of the second law on somewhat more rigorous foundations.

Novel or not, however, Zermelo's challenge prompted Boltzmann to develop his time-symmetric view further, and to address the two unwelcome consequences described above: the symmetry of the implication that entropy increases, and the mystery as to why it is low now. Boltzmann not only recognized these consequences more clearly than his contemporaries, but suggested an audacious way to respond to them. Taking the latter point first, he realized that it was a consequence of the statistical argument that although the low-entropy condition of the world is very unusual, it is also inevitable, given enough time. So in a sense what is unusual is not that there *exist* regions like ours—that in itself is inevitable, in a sufficiently large universe—but that we find ourselves in such a region.[13]

However, Boltzmann suggested that it is very plausible that life itself depends on the low-entropy condition, in such a way that intelligence could not exist in the regions of the universe which are not in this condition. Life on Earth depends on a continual flow of energy from the sun, for example, and would not be possible without a low-entropy hot-spot of this kind. In light of considerations of this kind, Boltzmann suggests that it isn't surprising that we find ourselves in a low-entropy region, rare as these might be in the universe as a whole. This is an early example of what has come to be called *anthropic* reasoning in cosmology. By way of comparison, we do not think that it is surprising that we happen to live on the surface of a planet, even though such locations are no doubt rather abnormal in the universe as a whole.

But what of the fact that we live in a region in which—by our lights, at least—entropy goes up rather than down? Boltzmann suggests that this too can be explained in terms of the conditions needed to support creatures like us. Specifically, he suggests that our sense of past and future depends on the entropy gradient, in such a way that we are bound to regard the future as being the direction in which entropy increases. Boltzmann backs this up by suggesting that elsewhere in the universe, or in the distant past or future, might be creatures with the opposite temporal orientation, living in regions in which the predominant entropy gradient goes the other way. These creatures would think of our past as their future, and our future as their past; and there wouldn't be a fact of the matter as to which of us was right. (As we shall see in chapter 4, this possibility also emerges in discussions of time asymmetry in contemporary cosmology.)

Boltzmann explains these points in terms of a simple characterization of the graph of an isolated system over time. He points out that the statistical argument implies that the system spends nearly all its time at, or very close to, the level of maximum entropy (or minimum H). Just occasionally, however, one of the inevitable statistical fluctuations takes the system to a state of significantly lower entropy—and the lower the entropy concerned, the rarer (by far) the fluctuation required. The effect of the last point is that for a given entropy level below the maximum, it is very probable that when the system is at that level, it is close to one of the minima of the curve, and hence is either already increasing or about to do so *in both temporal directions*. But for the creatures like ourselves who live on the slope of such a fluctuation, where the local curve is strongly asymmetric, the direction of increase is bound to be taken as the future, because of the way in which their own physiological processes depend on the entropy gradient.

This ingenious suggestion explains the apparent asymmetry of thermo-dynamics in terms of a cosmological hypothesis which is symmetric on the larger scale. Moreover, it clearly embodies the idea that the direction of time is not an objective matter, but an appearance, a product of our own orientation in time. In both respects Boltzmann's suggestion is a great advance not only on previous work, but also on that of most of his successors.

But is it plausible? Its difficulties are of two main kinds. There are intrinsic difficulties stemming from the statistical approach itself, and there are problems which stem from elsewhere in physics. The main intrinsic difficulty arises from Boltzmann's own recognition that the most probable way to achieve a given entropy below the maximum is to achieve it at, or very close to, a minimum of the graph of entropy over time. If we wish to accept that our own region is the product of "natural" evolution from a state of even lower entropy, therefore, we seem bound to accept that our region is far more improbable than it needs to be, given its present entropy. One might try to invoke the anthropic argument again at this point, arguing that other possible configurations of the present entropy level would not have the history required for the evolution of life. However, this seems to miss the point of the connection between entropy and probability. If the choice is between (1) fluctuations which create the very low-entropy conditions from which we take our world to have evolved, and (2) fluctuations which simply create it from scratch with its current macroscopic configuration, then choice (2) is overwhelmingly the more probable. Why? Simply *by definition,* once entropy is defined in terms of probabilities of microstates for given macrostates. So the most plausible hypothesis—overwhelmingly so—is that the historical evidence we take to support the former view is simply misleading, having itself been produced by the random fluctuation which produced our world in something very close to its current condition. (It is no use objecting that such a fluctuation would have to involve all kinds of "miraculous" correlated behavior. It would indeed, but not half as miraculous as that required by option [1]!)

So Boltzmann's suggestion seems to have the unfortunate consequence that it implies that our apparent historical evidence (including the state of our own memories!) is almost certainly unreliable. We haven't arrived at this point by sedate progression from some earlier state of lower entropy, in other words, but by a "miraculous" fluctuation from a state of higher entropy. Indeed, we can even say what the "miracle" should look like. The argument suggests that by symmetry, our past should be something like a mirror image of our future. The normal progression of our world toward equilibrium also represents the most probable path away from equilibrium, toward a world like ours. (Again, this follows from the fact that the statistical considerations involved are time-symmetric: they simply involve counting combinations, and don't pay any attention to the distinction between past and future.) So in statistical terms, the most reasonable hypothesis is that the actual history of our region—the actual progression *from* equilibrium, in our time sense—is more like a mirror image of this future progression to symmetry, than an evolution from a state of lower entropy.[14]

There is a second consequence of a similar kind. Just as the cheapest way to produce the observed low-entropy state of our own region is in a fluctuation which does not unnecessarily extend the low-entropy condition to earlier times, so too one should avoid fluctuations which extend the low-entropy region unnecessarily in space. In other words, it seems to follow from Boltzmann's suggestion that we should not expect the low-entropy region to be any more extensive than we already know it to be. Indeed, as in the historical case—perhaps it is really the same thing—we might do even better by taking most of our current "evidence" about distant parts of space to be misleading. It would be cheaper to produce pictures of galaxies than galaxies themselves, for example. But that aside, we should expect to find no more order than we already have reason to believe in. Astronomy seems to defeat this expectation, however. We now observe vastly more of the universe than was possible in Boltzmann's day, and yet order still extends as far as we can see.[15]

This brings us to another aspect of the case against Boltzmann's hypothesis. Twentieth-century cosmology seems to show that the universe is simply not old enough to produce the kind of massive statistical fluctuations which Boltzmann's suggestion requires. At the same time, as we shall see in chapter 4, it suggests an alternative account as to why entropy is low in what we call the past. So it suggests that Boltzmann's ingenious hypothesis is unnecessary, as well as implausible.

The lesson that the contemporary debate needs to learn from Boltzmann is this one: *the* major task of an account of thermodynamic asymmetry is to explain why the universe as we find it is so far from thermodynamic equilibrium, and was even more so in the past. (We should also bear in mind that we have found no reason to doubt Boltzmann's anthropic hypothesis—viz., that the fact that we regard the temporal direction concerned as the past is itself to be explained in terms of the low-entropy condition of that region.) However, there is a very important difference between the contemporary perspective and that embodied in Boltzmann's own suggestion. Boltzmann's idea was that the current low-entropy state of our region is a product of a rare but inevitable statistical fluctuation. Coupled with the anthropic argument as to why we *find ourselves* in such a region, this explanation needs nothing over and above the basic statistical premise: roughly, the assumption that all possible microstates are equally likely.

The above objections suggest that a purely statistical explanation of the low-entropy condition of the present and past universe is unsatisfactory, however. To the extent that there are viable contemporary alternatives, they therefore involve an extra ingredient. (Not all contemporary cosmologists appreciate this point, however, as we shall see in chapter 4.) If the low-entropy initial condition of the universe is to be satisfactorily explained, it needs to be argued not simply that it is just a very lucky accident, but that some additional constraint operates to ensure that the universe begins in this way. As we shall see, this seems to make a big difference to the legitimacy of applying the statistical argument in the other temporal direction, in an attempt to explain why entropy increases toward what we call the future.

DO WE NEED TO EXPLAIN WHY ENTROPY INCREASES?

We have credited Boltzmann with the recognition that one of the major puzzles of thermodynamics is the question as to why entropy is low *now*, and apparently even lower in the past. If equilibrium is the natural condition of matter, why is the matter around us so far from achieving it? With this issue in our sights, however, do we also need to explain why entropy increases?

At first sight the latter question may seem absurd. After all, the universal increase in entropy is the central asymmetry identified by thermodynamics. Isn't it obvious that it requires explanation? This response misses the point, however, for suppose we had succeeded in explaining why entropy is low in the past—i.e., in effect, why the entropy curve slopes downwards in that direction. Is there a separate question as to why it slopes upwards in the other direction, or is this just another way of asking the same thing?

Historically it seems to have been taken for granted that there is a separate issue here, though it needs to be borne in mind that not everyone had a clear grasp of the fact that the low-entropy past is itself in need of explanation. At any rate, the perceived problem is the one to which a suitably generalized version of Boltzmann's *H*-theorem would be a solution, in effect, if it could be shown not to involve a problematic asymmetric assumption. Such a result would provide an explanation for the general increase in entropy in all systems, in terms of the dynamical behavior of their constituent parts. The fact that Boltzmann himself sought to defend the *H*-theorem in light of the criticism of Culverwell, Burbury, and others in the 1890s suggests that he himself saw a need for an explanation of this kind. His critics agreed with him, of course. The point of Burbury's attempted appeal to the disturbing effects of external influences was supposed to have been to put Boltzmann's own argument on a sound footing. And while later writers might challenge the Boltzmann-Burbury approach to the *H*-theorem, they almost never challenge the basic project. Almost everyone seems to believe that we need an explanation as to why entropy always goes up.

But do we? What happens if we look at the problem in reverse? From this perspective it seems that entropy is high in the past, and always decreases. The universal tendency to decrease toward what we now take to be the future looks puzzling, of course, but might be taken to be adequately explained if we could show that the laws of physics impose a boundary condition in that direction. Why does entropy decrease? Because it is a consequence of certain laws that the entropy of the universe must be low at such and such a point in the future. Each individual decrease would thus be explained as a contribution to the required general decrease.

Do we need to explain why entropy is high in the past, in this picture? Apparently not, for according to the statistical account, that is not an exceptional way for the past to be. All we need to do, apparently, is to note that in that direction the universe does not appear to be subject to the boundary constraint that imposes low entropy toward the future. In other words, from this perspective the real work of explaining why entropy shows a universal tendency to decrease is done by the account of why it is low at a certain point in the future, together with the remark that the past is not similarly in need of explanation.

However, if we accept that this is a satisfactory account of what we would see if we looked at our universe in reverse, it is hard to maintain that it is not a satisfactory explanation of what we actually do see—for the difference between the two views lies in our perspective, not in the objective facts in need of explanation. And yet from the ordinary viewpoint all the work is done by the account of why entropy is low in the past. The future seems to need no more than a footnote to the effect that no such constraint appears to operate there, and that what we foresee in that direction is not in need of explanation, for it is the normal way for matter to be.

This conclusion applies to the countless individual processes in which entropy increases, as well as to the second law in general. Consider what happens when we remove the top from a bottle of beer, for example: pressurized gas escapes from the bottle. Traditionally it has been taken for granted that we need to explain why this happens, but I think this is a mistake. The gas escapes simply because its initial microstate is such that this is what happens when the bottle is opened. As the tradition recognizes, however, this isn't much of an explanation, for we now want to know *why* the initial microstate is of this kind. But it seems to me that the correct lesson of Boltzmann's statistical approach is that this kind of microstate doesn't need any explanation, for it is overwhelmingly the most natural condition for the system in question to possess. What does need to be explained is why the microstate of the gas is such that, looked at in reverse, the gas enters the bottle; for it is in this respect that the microstate is unusual. And in the ordinary time sense, this is just a matter of explaining how the gas comes to be in the bottle in the first place.

One point to notice here is that we ordinarily explain later conditions of physical systems in terms of earlier conditions. We have no particular reason to expect this asymmetry to survive when we move to an atemporal perspective. After all, it is a notion of explanation that we associate with the idea of one set of conditions "giving rise to" another, and yet the notion of "giving rise to" seems to presuppose a temporal direction. The issue as to what form explanation properly takes from the atemporal perspective is a big one, which I won't attempt to tackle here. However, at least two features of the ordinary conception of explanation seem likely to survive the move unscathed. One is the theoretical notion of explanatory priority, the idea that some things explain others in virtue of their more basic role in theory. The second is the notion that some features of reality are more exceptional or unusual than others, and therefore more in need of explanation.

Here the point connects with an observation we made at the beginning of the chapter. The history of science shows that science often changes our conception of what calls for explanation and what doesn't. Familiar phenomena come to be seen in a new light, and often as either more or less in need of explanation as a result. One crucial notion is that of normalcy, or naturalness. Roughly, things are more in need of explanation the more they depart from their natural condition, but of course science may change our view about what constitutes the natural condition. (The classic example is the change that Galileo and Newton brought about in our conception of natural motion.)

It seems to me that the lessons of the second law may be seen in this light. What the discussion in the second half of the nineteenth century revealed is that thermodynamic equilibrium is a natural condition of matter, and therefore that it is departures from this condition that call for explanation. Our world exhibits a huge and apparently monotonic such departure toward what we call the past, the explanation of which is the major task revealed by thermodynamics. Insofar as we have reason to think that what we call the future has high entropy, this does not call for explanation in the same sense, for the statistical considerations reveal that this is the natural condition of matter.[16]

Thus it seems to me that the problem of explaining why entropy increases has been vastly overrated. The statistical considerations suggest that a future in which entropy reaches its maximum is not in need of explanation; and yet that future, taken together with the low-entropy past, accounts for the general gradient. In sum, the puzzle is not about how the universe reaches a state of high entropy, but about how it comes to be starting from a low one. It is not about what appears in our time sense to be the destination of the great journey on which matter is engaged, but about the point from which—again in our time sense—that journey seems to start. (The shift in perspective we need here is something like that of the Irishman of comic fame, who was asked for directions to Galway: "If it's Galway you're after," he replied, "why the devil are you leaving from here?")[17]

THE ROLE OF THE *H*-THEOREM

The conclusion of the previous section will be controversial, for it flies in the face of a century-old endeavor in physics. We noted that when Boltzmann sought to defend the *H*-theorem in the 1890s, his critics—Culverwell, Burbury, and others—were not opposed to the *H*-theorem itself, but simply wanted to put it on sound foundations. And this is by far the dominant tone of later discussions. Later writers will often challenge Boltzmann's right to the *stoßzahlansatz,* pointing out for example that it is itself temporally asymmetric, but they see in this a profound difficulty for a worthwhile project. Almost everyone seems to believe that we need some such *dynamic* explanation for the fact that entropy always increases—in other words, that we need an explanation in terms of the "natural" behavior of the constituents of matter over time.

Where does this project stand in the light of the above discussion? My view is that as an assault on the fundamental issues of temporal asymmetry in thermodynamics, it simply misses the point. The basic problem is that the *H*-theorem has its eyes set to the future, while all the interesting action is actually in the past. The theorem is trying to explain something which the statistical considerations show not to be in need of explanation—while failing to address the important issue.

The fact that the project misses the point in this way is revealed in the attention given to the asymmetric presuppositions of the theorem—that is, to Boltzmann's *stoßzahlansatz,* or principles such as our PI[3]—as if these were the *origin* of the asymmetry of thermodynamics. In reality it is not these assumptions themselves that are puzzling, but the fact that they fail in reverse. After all, if they did not fail in reverse then the theorem could be applied in reverse, to "show" that entropy increases (or *H* decreases) toward the past, as well as toward the future. Since it is the fact that entropy does not behave in that way toward the past that actually requires explanation, it is the failure of the required assumptions in that direction that should be the focus of attention. A statistically "natural" world would be one in which the assumptions concerned held in both directions, not a world in which they held in neither.

More on the status of PI[3] and its relatives below. First let me qualify these remarks about the *H*-theorem in one important respect. Recognition that the general problem concerns the past rather than the future does not exclude a more local interest in the behavior of systems on the entropy gradient. It is one thing to characterize the endpoints of an entropy gradient, another to say how systems of one kind or another get from one end to the other. This local project is by definition a temporally asymmetric one. It concerns the behavior of systems assumed to possess an entropy gradient in a particular direction—systems constrained in one direction but not the other. It seems entirely appropriate to seek general *descriptions* of systems of this kind, and to see the *H*-theorem and its descendants in this light. On this reading, the task of the *H*-theorem is not to account for the existence of an entropy gradient, but to characterize the behavior of matter on the gradient, given that there is one.

This is a very different project from that of *explaining* the entropy increase, however. For one thing, the descriptive project is compatible with the atemporal perspective. We don't have to see the argument as having the force of a dynamical explanation—as describing what "pushes" the system in question toward high entropy. (As before, such an explanation simply misses the point about what needs to be explained.) Rather it enables us to describe the most probable path between the two endpoints: not the most probable path *from* A *to* B, or from B to A, but simply *between* A and B.

For another thing, the asymmetry of PI[3] and its relatives is not problematic when the *H*-theorem is seen in this descriptive sense. It simply has the status of a kind of default assumption, which we take to hold, other things being equal. If the description turned out wrong in a particular case, we would simply ask why the assumption failed to hold in that case.

How exactly should the required asymmetric assumption be characterized? Given the reduced explanatory demands on the *H*-theorem, this question does not seem as pressing as it does in the orthodox debate. This is fortunate, I think, for it is doubtful whether the assumption can be characterized in a noncircular way. After all, what we are looking for, essentially, is a necessary and sufficient condition on the constituents of matter for entropy to be guaranteed to increase. If the condition weren't sufficient then the explanation would be incomplete, and if it weren't necessary then we wouldn't have characterized the condition whose failure in the reverse time sense allows entropy to go down. But it is doubtful whether this leads to anything except the trivial condition that the constituents of matter be arranged in such a way that entropy increases. If we see the task as explanation, in other words, we find ourselves saying that entropy goes up because matter is arranged in a configuration such that entropy goes up.

To illustrate this point, it might be useful to consider the claim of a number of contemporary writers that the second law can be explained by the independence of initial conditions early in the history of the universe. In his 1987 book *Asymmetries in Time,* for example, the philosopher Paul Horwich suggests that it is the fact that the initial microstate of the universe is highly random that explains why entropy generally increases. He points out that in contrast the final microstate must be highly correlated, reflecting the fact that it is the deterministic product of a very highly ordered (low-entropy) initial state.

What does this proposal actually amount to, however? Note that it would seem completely inappropriate to say that the universe has a highly ordered initial state *because* it is later in a highly correlated microstate. If there is an adequate explanation in either temporal direction in this case, then it goes from past to future: the low-entropy early state explains the highly correlated microstate at later stages. Alternatively, and I think more accurately, we might say that there is no real explanation in either direction—we simply have two ways of describing the same fact about the universe, in effect. Either way, however, we don't take the *final* microstate to explain the *initial* ordered state.

But then by what right do we propose that an *initial* microstate can explain a *final* macrostate? In practice, of course, we are inclined simply to help ourselves to the principle that the past explains the future, but what could possibly justify that inclination here, where the temporal asymmetry of the universe is what we are seeking to explain? It seems that from the atemporal viewpoint we have no more right to take initial micro chaos to explain the later macrostate than we do to take the final microstate to explain the initial macrostate. Again, either it is the later macrostate which explains the earlier microstate, or—and this seems to me the more appropriate conclusion—there is no substantial explanation in either direction, so that we simply have two ways of describing the same phenomenon.

Thus it seems to me that the project of explaining entropy increase in terms of PI[3] and its relatives turns out to be vacuous, as well as misdirected. In the last analysis, the requirement that initial conditions must satisfy for entropy to increase toward the future is the requirement that final conditions must fail to satisfy, given that entropy decreases toward the past. In other words, in effect, it is just the requirement that the initial conditions not be correlated in such a way that entropy decreases in the future. There is no explanation in the offing here, but only another way of describing what was thought to require explanation.

The fact that PI^3 does not play the role it is often thought to play in accounting for the thermodynamic asymmetry will be important later in the book (see chapter 5), when we call into question the credentials of a related principle which is usually taken for granted in microphysics. To finish this chapter, I want to illustrate the conclusions of this and the previous section by outlining their impact on two more orthodox approaches to the asymmetry of thermodynamics; I want to show how these approaches look rather misdirected, once we understand that the real puzzle concerns the low-entropy past, and see the issue in atemporal terms. Also, to connect our present concerns with those of chapter 4, I want to consider the issue as to whether there might be a low-entropy future, as well a low-entropy past.

DOES CHAOS THEORY MAKE A DIFFERENCE?

In recent years it has often been suggested that the key to the apparent conflict between thermodynamics and mechanics lies in chaos theory, and the application of nonlinear methods in physics. This view is particularly associated with the Brussels School, led by the Nobel Prize–winning theoretical chemist Ilya Prigogine, known especially for his work on the behavior of systems which are far from thermodynamic equilibrium.

Like the traditional approach to the *H*-theorem, however, this suggestion rests on a failure to understand the real puzzle of thermodynamics. As I have emphasized, the puzzling asymmetry of thermodynamics consists in the fact that entropy is low in the first place, not in the fact that it later increases. The nonlinear approach may tell us how certain nonequilibrium systems behave, *given that there are some*—like the *H*-theorem, it may play an important role in characterizing the nature of systems on the entropy gradient—but it doesn't explain how the gradient comes to exist in the first place. This is the great mystery of the subject, and the theory of nonequilibrium systems simply doesn't touch it.

A more direct way to appreciate the inability of these new approaches to resolve the old puzzles of asymmetry in thermodynamics is to note that they too are vulnerable to a version of the reversibility objection. We saw earlier that in its second form, the underlying insight of the reversibility objection is that a symmetric theory is bound to have the same consequences in both temporal directions. A particularly powerful way to apply this insight is as follows.

Suppose that the proponents of the nonlinear dynamical method—or any other dynamical method, for that matter—claim that despite the fact that it is a symmetric theory, it produces asymmetric consequences in thermodynamics, and hence avoids the old paradoxes of reversibility. To undermine their claim, we describe an example of the kind of physical system to which the new method is supposed to apply, specifying its state at some time *t*. We then ask our opponents to tell us the state of the system at another time, say, *t* + 1, *without being told whether t* + 1 *is actually earlier or later than t*. (That is, without being told whether a positive time interval in our description corresponds to a later or an earlier time in the real world.) If our opponents are able to produce an answer without this extra information, then their theory must be time-symmetric, for it generates the same results in both temporal directions. If they need the extra information, on the other hand, this can only be because at some point their theory treats the two directions of time differently—like Boltzmann's original *H*-theorem, in effect, it slips in some asymmetry at the beginning. So in neither case do we get what the advocates of this approach call "symmetry-breaking": a temporal asymmetry which arises where there was none before. Either there is no temporal asymmetry at any stage, or it is there from the beginning.

This simple argument is really just a graphic way of making the point which underlies Culverwell's challenge to the *H*-theorem in the 1890s. No theory of the evolution of a physical system over time can produce different results for the two temporal directions, unless it treats them differently in the first place.[18]

For all their intrinsic interest, then, the new methods of nonlinear dynamics do not throw new light on the asymmetry of thermodynamics. Writers who suggest otherwise have failed to grasp the real puzzle of thermodynamics—Why is entropy low in the past?—and to see that no symmetric theory could possibly yield the kind of conclusions they claim to draw.

BRANCH SYSTEMS

Writers on the asymmetry of thermodynamics have often puzzled about the fact that the second law seems to govern not only the universe as a whole, but also its countless subsystems of various sizes. Many of these subsystems seem to be largely isolated for long periods of time. How is it that their direction of entropy change is nevertheless coordinated with that of other subsystems, and of the universe at large? Why are the individual "branch systems" all aligned in this way?[19]

One approach to these issues has to been to take up Burbury's suggestion that the second law depends on random external influences. It is argued that in practice no branch system is ever completely isolated from the external world, and hence free of the effects of such influences. Against this idea, however, some writers appeal to the intuition that entropy would still increase—the milk and the coffee would still mix, for example—even if the system were completely isolated. And as I have argued above, it is doubtful whether the required notion of randomness can be characterized in such a way as to make this approach nonvacuous.

Other writers appeal to randomness more directly, relying on PI^3, or something very like it. Once again, however, it is doubtful if this does more than to restate the problem these writers take themselves to be addressing, for it amounts to the assumption that the initial conditions in branch systems are always such that entropy increases. As the philosopher Lawrence Sklar puts it,

> This is just to repeat the fact that systems do show thermodynamic behavior (i.e., approach to equilibrium) in a parallel time direction and, indeed, in the direction of time we call the future. But the arguments ... don't provide a physical explanation for that fact. Rather they once again build it into their description of the world as a posit.[20]

Sklar goes on to deny that an appeal to a low-entropy initial condition for the universe is of any use at this point: "Cosmology by itself, including the Big Bang [and] its low entropy ... doesn't seem to provide the explanation of the parallelism in time of the entropic increase of branch systems." But a low-entropy initial condition does address the problem of parallelism, so long as we pose the issue in reverse: Why does entropy in branch systems always decrease toward the past? Simply because all (or very nearly all) the low-entropy branch systems we encounter ultimately owe their condition to the state of the universe very early in its history. They have all escaped from the same bottle, in effect, and this explains why they all approach that bottle as we follow them toward the past. (More on the details of this cosmological account in chapter 4.)

Is there a separate issue as to why all the systems are moving away from the bottle, as we look toward the future? No, because this is just another way of describing the same thing. There is no separate problem as to why entropy in branch systems always increases toward the future, in other words: only the big problem as to why everything was in the bottle in the first place.

It is true that once we realize the importance of a low-entropy constraint in the past, it is natural to ask why there doesn't seem to be a corresponding constraint in the future. This can be phrased as the question why we don't find the kind of entropy-reducing systems that such a future constraint would entail, but it is not really the same issue as that traditionally addressed under the heading "Why does entropy always increase?" The answer it seeks lies in cosmological constraints on the universe at large, not in some dynamical constraint on the behavior of individual systems. It doesn't challenge what I take to be the crucial lesson of the statistical approach to thermodynamics, namely that high-entropy states are not unusual, and therefore don't need explaining.

COULD ENTROPY EVENTUALLY DECREASE?

The suggestion just mentioned brings us to an issue which connects the concerns of this chapter with those of chapter 4. Is it really a possibility that there might be a constraint in virtue of which entropy will eventually decrease toward the future, like the constraint in virtue of which it decreases toward the past?

At first sight it might seem that statistical considerations rule this out. Wouldn't such a constraint require that the future universe be in an immensely unlikely condition? It would, but the significance of this fact seems to be very sensitive to the kind of the explanation we offer for the low-entropy past. We saw earlier that according to Boltzmann's own suggestion, the low-entropy past is itself explained by the basic statistical hypothesis—by the assumption that all microstates are equally likely. On this view, then, the existence of a region of the universe in this statistically exceptional condition does not threaten the statistical premise itself; it does not suggest that the statistical hypothesis is false. (It is the kind of rare "counterexample" the hypothesis itself leads us to expect, in other words.) On the other hand, Boltzmann's proposal also implies that entropy is bound to decrease in the future, at least if the universe lasts long enough. However, the timescales involved are such that we could be very confident that there is no such low-entropy future in any cosmological region of any relevance to us.

If Boltzmann's proposal is rejected, however, and we seek to explain the low-entropy past in terms of some additional assumption or principle, then we have introduced something that conflicts with the hypothesis that all microstates are equally likely. The introduction of a new factor of this kind has a major bearing on whether the statistical argument gives us reason to rule out the possibility that the thermodynamic arrow will reverse. For the statistical argument is a weapon with two possible uses—as Boltzmann himself realized, the argument can be "pointed" in either temporal direction. However, in accepting that an additional principle is needed to explain the low-entropy past, we accept that the statistical argument is unreliable in that direction. It encounters a counterexample (a genuine one, rather than the kind of rare "fluctuation" it leads us to expect), and the statistical considerations are "trumped" by the additional constraint.

If we accept that the statistical argument doesn't work in one of the two temporal directions, should we trust it in the other direction? Apparently not, for we have no independent means of checking its credentials, and its twin is a failure in the only comparable case. By way of comparison, suppose that we are offered similar cars by used car salesmen Keen and Able, who happen to be identical twins. They both say, "Trust me, this car is completely reliable." We try Keen's car, but the wheels fall off before we leave the yard. Should we trust Able? Not unless we have some independent reason for thinking that the two cases are different.

What would such a reason amount to in the temporal case? It would be a reason for thinking that the future is *not* subject to the kind of constraint that ensures that entropy is low in the past. In the absence of such a reason I think we are guilty of an unjustified *double standard* if we rely on the statistical argument in one direction but not the other. The best we can say seems to be something like this: the statistical argument gives us reason to expect that entropy will increase, *unless it is constrained to do otherwise.* This isn't quite a matter of "entropy will go up unless it goes down," but only because the high-entropy state continues to be the default—the normal condition, which we don't need to explain. It is a lot weaker than the conclusion the statistical argument is normally thought to supply. In particular, it is absolutely powerless to rule out the possibility that the universe is symmetric, in the sense that whatever requires that entropy be low in the past also requires that entropy be low in the future. As we shall see in chapter 4, contemporary cosmology has had considerable difficulty grasping this point, and double standards of this kind remain very common.

SUMMARY

1. We need to be careful to avoid unrecognized asymmetric assumptions. Recognized asymmetric assumptions are acceptable, as long as we see that they simply shift the explanatory burden from one place to another. The tendency to make this mistake in trying to explain the asymmetry of thermodynamics has been quite well recognized over the years, but its true significance has been missed. It has not been appreciated that the problem stems from a misunderstanding about what really needs to be explained about the thermodynamic arrow. With a clear view of the explanatory task (see [2] below), the project that runs into this problem—that of explaining why entropy *increases*—can be seen to be misconceived.

2. The real puzzle is to explain the low-entropy past. The standard debate has tended to concentrate on the wrong end of the entropy gradient. Writers have asked about the assumptions required to show that entropy increases

toward (what we call) the future, for example, rather than the conditions required to ensure that it decreases toward (what we call) the past.

3. At the root of this mistake lies a failure to characterize the issue in sufficiently atemporal terms. Despite Boltzmann's progress toward a view of this kind in the 1870s and 1890s, it has not been properly appreciated that we have no right to assume that it is an objective matter that entropy *increases* rather than *decreases*, for example. What is objective is that there is an entropy gradient over time, not that the universe "moves" on this gradient in one direction rather than the other.

4. We need to be on our guard against temporal double standards—that is, against the mistake of accepting arguments with respect to one temporal direction which we wouldn't accept with respect to the other. Double standards need to be justified, and since a justification has to provide a reason for treating the past and the future differently, it is bound to embody or rely on some further principle of temporal asymmetry (so that conclusion [1] now applies).

5. The plausible but asymmetric principle we called PI^3, which seemed to underpin Boltzmann's *stoßzahlansatz*, or assumption of molecular chaos, turns out to have surprisingly little role to play. The project that seemed to require it turns out to be misconceived, and the principle itself seemed difficult to formulate in nonvacuous terms (i.e., except as the principle that things are not arranged in such a way that entropy decreases). As a powerful intuitive temporal asymmetry, however, it is an item that we should flag for future consideration. Later in the book we shall see that a close relative of PI^3 plays a powerful role elsewhere in physics, but that its credentials are rather dubious, and that we might do better to abandon it.

3

←——→

New Light on the Arrow of Radiation

In his book *Ring of Bright Water,* the writer Gavin Maxwell describes his life with an otter, Mijbil, on the northwest coast of Scotland. If we ignore the otter and think about the water, we come to the problem I want to talk about in this chapter. Most of us have never seen an otter slip into a still Scottish loch, but we all know what happens, at least in one respect. Circular ripples spread outwards from the otter's point of entry, across the water's surface. It turns out that this familiar and appealing image illustrates another of the puzzling ways in which nature is asymmetric in time—and the asymmetry involved turns out to be different, at least at first sight, than the kind we discussed in the previous chapter.

The asymmetry turns on the fact that the ripples always spread *outwards,* rather than *inwards.* Ripples spreading inwards would be just the temporal inverse of ripples spreading outwards—they are what we see if we play a film of the otter in reverse, after all—but in nature we seem to find only ripples of the outgoing kind. (It makes no difference whether the otter is entering or leaving the water, of course!) The same thing turns out to be true not merely of disturbances in water caused by things other than otters, but also, more surprisingly, of all other kinds of wave-producing phenomena in the physical world. The resulting asymmetry has become known as the asymmetry (or "arrow") of radiation. It is a feature of a very wide range of physical processes involving the radiation of energy in the form of waves. In a recent book on the physics of time asymmetry, Dieter Zeh describes it like this:

> *After* a stone has been dropped into a pond one observes concentrically *out*going waves. Similarly, *after* an electric current has been switched on, one finds a retarded [i.e., outgoing] electromagnetic field. Since the laws of nature which successfully describe these events are invariant under time-reversal, they

49

are equally compatible with the reversed phenomena in which, for example, concentrically focussing waves would eject a stone from the water. ... Such "reversed processes" have however never been observed in nature.[1]

If ingoing and outgoing waves are equally compatible with the underlying laws of physics, why does nature show such a preference for one case rather than the other? This is the puzzle of the arrow of radiation, and its discovery by physicists provides another illustration of the very selective gaze with which science looks at the world. As I noted in the previous chapter, we can't "see" asymmetry in the world until we have a conception of what symmetry would look like. Although water waves are familiar to everybody, it seems to have been the case of electromagnetic radiation—light, radio waves, and the like—that first brought the asymmetry of radiation to the attention of physics. As in the case of thermodynamics, Maxwell (James Clerk, not Gavin) again played a crucial role. Maxwell's theory of electromagnetism, developed in the mid–nineteenth century, is easily seen to admit two kinds of mathematical solutions for the equations describing radiation of energy in the electromagnetic field. One sort of solution, called the *retarded* solution, seems to correspond to what we actually observe in nature, which is outgoing concentric waves. The other case, the so-called *advanced* solution, describes the temporal inverse phenomenon—incoming concentric waves—which seem never to be found in nature.

Thus the puzzle of temporal asymmetry here takes a particularly sharp form. Maxwell's theory clearly permits both kinds of solution, but nature appears to choose only one. In nature it seems that radiation is always retarded rather than advanced. Why should this be so? And what is the relation, if any, between this temporal asymmetry and that of thermodynamics?

There was considerable interest in these issues in the early years of this century. The Swiss physicist Walther Ritz, a colleague and contemporary of Einstein, proposed that the retarded nature of radiation should be regarded as a fundamental law of nature, and suggested that such a law might be used to *explain* the thermodynamical arrow of time. In a famous exchange with Ritz, however, Einstein argued that the asymmetry of radiation was "exclusively due to reasons of probability,"[2] and therefore of the same origin as the thermodynamic asymmetry. His view was that neither is a fundamental law. Instead, both rest on what the initial conditions happen to be like in our region of the universe.

As Zeh points out, however, one disadvantage of Ritz's view is that a fundamental law to account for the asymmetry of electromagnetic radiation would not explain the asymmetry of other kinds of radiation, such as sound and water waves.[3] The use of the example of water waves in this debate is particularly associated with the philosopher Karl Popper. In a famous letter to *Nature* in 1956, Popper criticized what he took to be the commonly accepted view among physicists that thermodynamics provides the only significant kind of temporal asymmetry in the physical world. As he puts it,

> It is widely believed that all irreversible mechanical processes involve an increase in entropy, and that "classical" (that is, nonstatistical) mechanics, of continuous media as well as of particles, can describe physical processes only in so far as they are reversible in time. This means that a film taken of a classical process should be reversible, in the sense that, if put into a projector with the last picture first, it should again yield a possible classical process.

Popper goes on to argue that this common view is mistaken:

> This is a myth, however, as a trivial counter example will show. Suppose a film is taken of a large surface of water initially at rest into which a stone is dropped. The reversed film will show contracting circular waves of increasing amplitude. Moreover, immediately behind the highest wave crest, a circular region of undisturbed water will close in toward the centre. This cannot be regarded as a possible classical process. (It would demand a vast number of distant coherent generators of waves the coordination of which, to be explicable, would have to be shown, in the film, as originating from one centre. This, however, raises precisely the same difficulty again, if we try to reverse the amended film.)

He concludes that

> irreversible classical processes exist. ... Although the arrow of time is not implied by the fundamental equations, it nevertheless characterizes most solutions. For example, in applying Maxwell's theory, the initial conditions determine whether we choose to work with retarded or with advanced potentials, and the resulting situation is, in general, irreversible.[4]

I'll criticize one aspect of Popper's argument below. For the moment, note that his view seems to be that the asymmetry of radiation is essentially independent of that of thermodynamics. This view is compatible with the hypothesis that the two "arrows" share a common explanation, either in terms of the initial conditions of the universe, or in terms of some common "master arrow," such as the expansion of the universe. However, it is incompatible with Ritz's suggestion that the thermodynamic asymmetry is a consequence of the radiative asymmetry. It is also incompatible with the more popular view—the direct opposite of Ritz's—that the arrow of radiation is a consequence of that of thermodynamics.

In this chapter I am going to argue for a version of the middle position. In other words, I want to show that the temporal arrow of radiation neither explains nor is explained by the arrow of thermodynamics. Instead it is a different species of the same genus, for the genealogy of which we need to look to the early history of the universe. Except for the modern cosmology, this seem to have been Einstein's view, as well as Popper's.[5]

However, it is important to distinguish two possible versions of this middle position. Earlier we characterized the asymmetry of radiation in terms of nature's apparent preference for outgoing ("retarded") rather than incoming ("advanced") solutions of the equations governing radiative phenomena. (This characterization seems as apt for other kinds of radiation as it is for the electromagnetic case.) One possible view would be that there really is no incoming radiation in nature, and hence that this absence of incoming radiation is what needs to be explained by boundary conditions (by factors in the early universe, for example). The other possible view is that there actually is both incoming and outgoing radiation in nature as we find it, but that in virtue of the way in which it is arranged—this arrangement to be explained by cosmological factors, perhaps—we simply don't notice the incoming radiation.

A simple example may help to clarify the difference between these two views. Suppose we have a chemical substance whose molecules may exist in either left-handed or right-handed forms. (Many real compounds are like this, including ordinary sugar.) Suppose we examine the molecules in a particular sample of the substance concerned, and find that all of those examined are of the left-handed variety. There are two possible explanations of this result. The first—analogous to the view that there really is no incoming radiation in the universe—is that the sample really contains only left-handed molecules. The second—analogous to the view that there is incoming radiation, but we simply don't see it—is that while there are both left-handed and right-handed molecules in the sample, we only see the left-handed kind. This might be due to some kind of bias in the observation technique. The right-handed molecules might be very soluble, for example. If we were examining the crystals found in our sample, this would explain why we only encountered the left-handed variety of the substance in question.

I want to defend a view of the latter kind about radiation. I want to argue that radiative processes are symmetric in themselves, and don't exhibit a preference for retarded solutions to the relevant wave equations—but that other factors make the retarded solutions much more noticeable.

My argument for this view is very unconventional. It involves a complete reinterpretation of a classic argument for the same conclusion by two of the giants of twentieth-century physics, John Wheeler and Richard Feynman.

Wheeler and Feynman's Absorber Theory of Radiation, developed when Feynman was Wheeler's student in the early 1940s, shares with my view the goal of showing that radiation is not intrinsically asymmetric, but "just looks that way." Wheeler and Feynman attempt to explain why radiation "looks asymmetric" by appealing to the thermodynamic asymmetry, so that in this respect they favor the view that the asymmetry of radiation is derived from that of thermodynamics, rather than the middle view that the two asymmetries are independent.

As I shall show, Wheeler and Feynman's argument is fallacious as it stands, both in its appeal to thermodynamics and in its conception of what intrinsically symmetric radiation would be like. Happily, however, it turns out that the mathematical core of the theory can be reinterpreted. In its new form it does indeed establish that radiation is fundamentally symmetric. The apparent asymmetry—the fact that we "see" only outgoing radiation, in effect—now turns out to be explained not *by means of* the thermodynamic arrow but in parallel with it, in the sense that both seem to depend on the same kind of cosmological boundary conditions.

The reinterpretation of Wheeler and Feynman's Absorber Theory has other advantages as well. We shall see that it avoids an ad hoc and physically implausible assumption required by Wheeler and Feynman's own version of argument. Also, it seems to free electromagnetism from some problematic cosmological constraints imposed by the standard Absorber Theory.[6]

In order to motivate the proposed reinterpretation I want to pay close attention to the question as to what the apparent temporal asymmetry actually involves. What exactly *is* asymmetric about radiation in the real world? It turns out that there are several different characterizations in common use. All seem *intended* to amount to the same thing—no one is proposing that radiation might exhibit two distinct temporal asymmetries, or professes to be in any doubt as to what the single asymmetry involves. However, we'll see that it is easy to be misled by certain of the common formulations, and so to misunderstand what the world would have to be like for radiation to be symmetric in time. I suspect that the simple solution proposed in this chapter has been obscured, in part, by a long-standing failure to pose the problem in precisely the right way.

Another goal of the chapter is to expose some common mistakes in discussions of the arrow of radiation—mistakes very similar to those we have seen to characterize discussion of the asymmetry of thermodynamics. In particular, many writers on the radiative asymmetry have been guilty of what I called the temporal double standard. They happily accept arguments with respect to one temporal direction which they would be unwilling to accept with respect to the other, without offering any justification for the difference. As we shall see in a moment, Popper himself is one of the culprits here, though the mistake is also characteristic of attempts to explain the asymmetry of radiation in terms of that of thermodynamics. (We shall also see that in the latter cases the double standard fallacy is in a sense even more serious than it is in the context of attempts to explain the second law in statistical terms, for it comes closer to logical circularity.)

There are two main arguments purporting to show that the retarded nature of radiation is a consequence of thermodynamics. One of them occurs in the context of the Wheeler-Feynman Absorber Theory, and we shall see later in the chapter that one of the failings of the standard version of the Absorber Theory is that it does commit the double standard fallacy in a particularly blatant way. The other argument is much simpler and much more common. Properly handled, however, its lessons are already sufficient to direct us to the main conclusions of the chapter.

THE CIRCULAR WAVE ARGUMENT

The simpler of the two arguments claiming to show that the retarded nature of radiation is a consequence of thermodynamics is easily illustrated in terms of Popper's pond example. Indeed, it is very closely related to the argument that Popper himself seems to favor. We want to know why still ponds are never observed to produce converging circular waves, let alone converging waves which arrive at their common center at just the right moment to give an added impetus to a stone recently miraculously expelled from its resting place on the bottom of the pond. The suggestion is that we consider what would be necessary at the edges of the pond (to say nothing of the bottom) in order for this to take place. At each spot the random motion of the edging material would have to cooperate to give the right sort of "nudge" to the adjacent water; and all these nudges would have to be precisely coordinated, one with another, at the different points around the pond. (In other words, they would have to occur simultaneously to give a wave converging to the center of a circular pond; and at appropriate temporal intervals, in any other case.)

At this point Popper himself assumes that such cooperative behavior is simply impossible (unless it is actually organized from a common center, which wouldn't be the reverse of the familiar case). If by "impossible" he simply means that it never happens in practice in the world as we know it, then of course he is right, but hasn't offered us any *explanation* for the temporal asymmetry: he hasn't told us *why* the world is asymmetric in this way. On the other hand, if "impossible" is intended in a stronger sense—in particular, if it is intended to support an *explanation* of the asymmetry—then Popper's argument involves the double standard fallacy. For if we look at the ordinary world in reverse we see exactly this kind of cooperative behavior. Played in reverse, a film of ordinary outgoing water waves shows exactly the kind of apparently collaborative behavior that Popper describes as impossible.

Looking toward what we call the past, then, this kind of behavior is not merely possible but exceedingly common. Anyone who wants to argue that it is impossible toward what we call the future had better be prepared to tell us what distinguishes the two cases. In other words, they had better be prepared to say why a principle that fails so blatantly in one direction is nevertheless reliable in the other. Of course, any adequate reason for distinguishing the two cases will be temporally asymmetric in itself, so that its effect will be to shift the bump in the carpet from one place to another, as I put it earlier. But this might amount to progress, even if it only enabled us to subsume the puzzle of the asymmetry of radiation under some more general heading. Popper's own argument doesn't get us this far, however, for he simply fails to notice the temporal double standard.

The argument takes a very similar form in the hands of those who see it as explaining the radiative asymmetry in terms of that of thermodynamics. As Paul Davies puts it,

> waves on a real pond are usually damped away at the edges by frictional effects. The reverse process, in which the spontaneous motion of the particles at the edges combine favourably to bring about the generation of a disturbance is overwhelmingly improbable, though not impossible, on thermodynamic grounds.[7]

Although the argument is formulated here in terms of water waves, the electromagnetic case is exactly parallel, so long as there is an absorbing boundary corresponding to the edge of Popper's pond. As Davies says, the point applies "quite generally to all types of waves in *finite* systems." Zeh also describes the electromagnetic version of the argument, concluding that "in this [finite] situation, the radiation arrow may thus very easily be derived from the thermodynamical one."[8]

Is the argument any more successful in this form that it was for Popper? Can it be regarded as explaining why we don't find converging concentric waves of water, sound, and light? The answer depends on the status of the second law of thermodynamics. If the second law were an exceptionless universal principle—as we saw it was originally thought to be—the argument would be a good one, I think, for it would simply amount to pointing out that in finite systems, converging radiation would require violations of this general law.

The argument is invalidated by the move to a statistical understanding of the second law, however. The mistake it now involves is essentially the same as that of Popper's argument above, and some of the mistakes we encountered in the last chapter. With a statistical understanding of the second law in place, it is being argued, in effect, that advanced radiation does not occur because it requires improbable behavior in surrounding materials. As we saw, however, these statistical considerations are symmetric. As Zeh himself puts the point:

> Trying to explain the situation by the remark that the advanced solutions would require *improbable* initial conditions would be analogous to the arguments frequently used in statistical mechanics. ... [T]he phenomena observed in nature are precisely as improbable.[9]

Thus the abnormal case of advanced radiation and the normal case of retarded radiation both involve events in surrounding matter which are overwhelmingly unlikely on statistical grounds alone. The coordinated events at the edges of ordinary (retarded) Popperian ponds occur *despite* their statistical improbability. The coordinated events that would be required in anti-Popperian ponds do not occur not simply because they are statistically improbable—that is true in the ordinary case as well—but also, crucially, because this statistical handicap is not overridden by the kind of favorable circumstances that allow the retarded case. But what would be a favorable circumstance? Why, simply the occurrence in the future of the kinds of events that *require* such coordinated predecessors: in other words, in particular, the occurrence of converging advanced waves, centered on disturbances such as outgoing stones. So in order to distinguish the normal and abnormal cases, the argument effectively *assumes* the very asymmetry it is trying to explain. It assumes that in nature we do have diverging radiation and don't have converging radiation!

As before, the easiest way to see the point is to adopt the atemporal perspective, or simply to imagine viewing things in reverse. From the reverse of the normal temporal perspective, ordinary water waves look incoming rather than outgoing. Why don't we take the occurrence of such waves to be ruled out by the improbability of the conditions at the edge of the pond required to give rise to them? Because these improbable conditions are *explained* by what now appear to be *later* circumstances at the center of the pond. (Remember that these are the circumstances we took to explain the waves concerned when we were looking at the phenomena in the usual temporal sense. Since we are assuming that in changing the perspective we don't change anything objective, we had better be consistent, and accept the explanation either in both cases or in neither.) Changing the perspective in this way helps us to see that the improbability argument cannot explain why there are no advanced waves (in the usual time sense). It shows us that there would be advanced waves, despite the improbability, if conditions at the center were as they are when we look at the "normal" case in reverse: in other words, if wave crests were converging, stones being expelled, and so on. The normal case shows that the statistical argument does not exclude these things as we look toward (what we call) the past. To take it to exclude them as we look toward (what we call) the future is thus to apply a double standard, and to fall into a logical circle—to *assume* what we are trying to *prove*.

As in the thermodynamic case, failing to notice the double standard means that we miss the real lesson of the statistical considerations. If we think of outgoing waves as normal, or "natural," and don't notice the vast improbability of the conditions they require in surrounding matter, we miss the real puzzle: Why are conditions so exceptional in (what we call) the past? In particular, why does the universe contain the kinds of events and processes which provide the *sources* for the outgoing radiation we observe around us? Why does it contain stars, radio transmitters, otters slipping into ponds, and so on? These kinds of things seem ordinary enough, and not collectively in pressing need of explanation, until we see that they have in common the fact that they give rise to a phenomenon which is in need of explanation, namely the kind of organized (or "coherent") outgoing radiation with which we are familiar.

Once we notice the important question in this way, I think it is immediately plausible that the search for an answer will lead us in the same direction as the issue as to why entropy is low in the past. Just as in the entropy case, we are led to investigate the provenance of a large range of different kinds of phenomena, which have a very general characteristic in common. Indeed, it seems plausible that the phenomena of relevance to radiation are simply a subclass of those of relevance to the second law. The general issue in the thermodynamic case is why energy exists in the past in such concentrated forms, and radiative sources are simply a special kind of concentration of energy—a kind in which dissipation takes a special form. Already then we have the outlines of an argument for what seems to have been Einstein's view of the relationship between the arrows of radiation and thermodynamics. The asymmetry of radiation seems to be neither a consequence nor a cause of the second law, but a phenomenon whose origins lie in the same puzzling features of the distant past.[10]

Our main task for the rest of the chapter is to confirm this diagnosis, and to show that it is best understood in terms of the second of the two possible versions of the Einstein view I distinguished in the previous section:

the version which says that there really is both incoming and outgoing radiation in nature, though only the latter kind is noticeable. Remember that I compared to this the idea that a chemical solution might contain both left-handed and right-handed versions of a molecule, even though only the left-handed variety is noticeable.

The first step is to clarify the problem. In order to avoid some confusions which tend to creep into discussions of these issues, we need a sharper understanding of what the asymmetry of radiation actually involves.

RADIATION AND BANKING

Why does a stone in a pond or an accelerated charged particle produce retarded rather than advanced radiation—a wave front which diverges into the future, rather than converging from the past? This question has much in common with the following one: Why does money deposited in my bank account on Wednesday appear in my balance on Thursday but not on Tuesday? There is one crucial difference between the two cases, which has to do with the fact that radiation is a dispersive phenomenon—the energy "deposited" in the electromagnetic field by a radio transmitter propagates away through space in all directions. I shall come back to this difference, for it is in this respect, if any, that radiation is more asymmetric than banking (and hence in this respect, I want to argue, that the reinterpreted Absorber Theory may be used to show that there is really nothing intrinsically asymmetric about radiation). Leaving aside this difference for the moment, however, I want to use the banking analogy to correct some misconceptions about the asymmetry of radiation.

There is nothing particularly puzzling about the temporal asymmetry of bank deposits. It is not the product of any intrinsic asymmetry in the activity of banking, but simply follows from what we mean by the term "deposit." A deposit just *is* a transaction in which a sum of money is *added* to the *prior* balance, thereby increasing the *subsequent* balance. The italicized terms all presuppose a temporal orientation. If we don't know whether a film taken in a bank is being projected forwards or backwards, we can't tell whether the masked figure it portrays is a thief or a rather eccentric depositor. What appears as a deposit from one temporal orientation appears as a withdrawal from the other, and vice versa. However, this dependence of terminology on temporal orientation is clearly extrinsic to the nature of banking itself. It doesn't mean that banking is intrinsically asymmetric in time.

Holding fixed the conventional temporal orientation, we may observe the following symmetry between deposits and withdrawals: the sum transferred appears in the account balance *after* a deposit but *before* a withdrawal. Again, there's nothing mysterious about this. On the contrary, it reflects an evident symmetry in the process of banking itself, namely the sense in which a withdrawal may be thought of as the temporal inverse of a deposit. In light of this symmetry, however, it may seem mysterious that there is another sense in which a withdrawal is not simply a deposit in reverse. For when do the *effects* of a deposit and a withdrawal manifest themselves in the banking system? In both cases, obviously, it is *after* the transaction takes place. The effect of a withdrawal is a reduction in one's balance; and this shows up after but not before the time of the transaction. Hence we might be tempted to say that banking turns out to be retarded rather than advanced—that although the opposite orientation is not ruled out by the laws of nature, in practice all banking transactions have an impact in the future and not the past.

However, I hope it is clear in the case of banking that it would be a mistake to locate the asymmetry in the mechanisms of banking itself. In particular, it would be a mistake to think that there is some alternative structure that banking could have had, but turns out not to have. The asymmetry is somehow a product of the way in which we apply the notions of cause and effect. Just what this "somehow" amounts to is a nice philosophical issue, to which we shall turn in chapters 6 and 7. Whatever the answer, however, the asymmetry has nothing particularly to do with the intrinsic processes of banking itself. These are as reversible as arithmetical operations of addition and subtraction. In fact, to all intents and purposes they are these operations, together with a means of maintaining a constant balance over the temporal intervals between transactions.

To what extent may we apply the lessons of the banking case to that of radiation? The obvious suggestion is that we should compare deposits with transmitters, which transfer energy *to* the electromagnetic field (or some other medium of wave propagation); and withdrawals with receivers, which transfer energy *from* the field. Notice that again this description assumes a temporal orientation—from the opposite orientation the flow of energy will appear reversed—but again this dependence on temporal orientation does not reflect or give rise to any intrinsic asymmetry in the processes themselves. The comparison also holds up in the sense that, concentrating on the energy balance alone (i.e., ignoring the dissipative aspects of the phenomena), reception is the temporal inverse of transmission. And finally, the parallel also holds in the sense that the effects of transmitters and receivers on the energy balance of the field both show up *after* the time of transmission or reception.

In the banking case we saw that this last feature does not show that banking is intrinsically retarded rather than advanced. The same is true for radiation. Whatever content there may be in the claim that radiation in nature is temporally asymmetric in being retarded but not advanced, it doesn't lie in the fact that both receivers and transmitters have retarded effects. On this point I disagree with Dieter Zeh, who appears[11] to take the fact that the effects of a receiver are delayed to refute the suggestion that radiation is intrinsically symmetric, reception simply being the inverse of transmission. The next section deals with a refinement in Zeh's argument.

RADIATION AND NONFRICTIONLESS BANKING

I have suggested that we compare emitters of radiation to bank deposits and receivers (or absorbers) of radiation to bank withdrawals. Leaving aside the dispersive aspect of radiation, the transfer of energy to and from the radiative field is no more asymmetric than the transfer of money in banking.

Moreover, it seems to me that this conclusion is not undermined if, as may well be the case, there are no pure receivers in nature; that is, if all receivers reradiate some of the incoming energy to the field. Zeh mentions this point in his initial characterization of the radiative arrow, and again in reply to the suggestion that the advanced solutions of Maxwell's equation simply characterize absorption, and therefore do exist in nature.[12] But it would be easy to construct banking systems with the analogous feature. Suppose that there are no overdrafts, so that withdrawals are only permitted up to the amount of one's current balance; and that the bank insists on redepositing 25 percent of any withdrawal. We might call this "nonfrictionless banking." Its result is that complete withdrawal of one's funds is possible only asymptotically, in the limit at infinity of an endless sequence of transactions.

Clearly the asymmetry of nonfrictionless banking is very superficial. The system may be described in the original symmetric terms, using the mirror-image notions of a pure deposit and a pure withdrawal, provided we add the stipulation that all actual withdrawals are "impure"—that is, they are a mixture of a pure withdrawal and a proportional pure deposit. In the case of radiation, the analogous idea would be that the advanced solutions do indeed characterize the radiation arriving at an absorber, but that real absorbers always retransmit a proportion of the energy they take from the field, so that their complete interaction with the field comprises a mixture of advanced and retarded solutions. This position seems adequate to characterize the phenomena that Zeh describes, and yet preserves the idea that radiation is intrinsically symmetric, in the sense that the advanced solutions do actually occur in nature.

Moreover, as the relevant proportion is reduced toward zero, the banking system in question approaches the original frictionless case; and this alone suggests that the impurity of receivers cannot account for the "all or nothing" asymmetry supposedly displayed by radiation in the real world. (We get much the same conclusion if we reflect on the quantized nature of energy transfer in the real world. At the microscopic level it must be possible for an absorber to take energy from the field without immediately reradiating it.)

WHAT WOULD TIME-SYMMETRIC RADIATION LOOK LIKE?

Thus it appears that if there is any substance to the standard claim that radiation is temporally asymmetric, we should look for it in the dissipative characteristics of the phenomenon; for it is only these that seem to afford any relevant distinction between radiation and banking. On the face of it, the relevant characteristic may seem obvious. A receiver may be the temporal inverse of a transmitter from the point of view of the energy balance, but surely there is this crucial difference: only a transmitter is centered on a coherent wave front; the waves incident on receivers are centered on transmitters elsewhere. This means that if we were shown a film depicting radiation, and could see the waves themselves, then we could tell whether what was depicted was transmission or reception (in the usual temporal orientation), without first knowing whether the film was being projected forwards or backwards. Indeed we could tell whether the film was running forwards or backwards, by noting whether concentric wave fronts appeared to be diverging or converging, respectively. Isn't this the crucial difference between transmitters and receivers, in virtue of which radiation may be said to be intrinsically asymmetric?

It seems to me that this apparent difference does reflect the most common understanding of the doctrine that radiation is retarded but not advanced. However, I have deliberately approached it indirectly, via the banking analogy. I wanted to ensure that when this difference turns out to be quite superficial—as I shall argue below that it is—it cannot be objected that I have misinterpreted the doctrine, and that there is some more problematic version waiting in the wings.

Even at this point, however, it turns out that there is room for confusion about the nature of radiative asymmetry. We have agreed that the temporal asymmetry of radiation consists in the fact that *emitters* but not *absorbers* are centered on coherent or organized wave patterns. In other words, a coherent pattern of waves radiates outwards from a radio transmitter, or from a stone dropped into a still pond; but it doesn't seem to be the case that coherent circular waves radiate inwards to a radio receiver, or to an energy-absorbing wave damper on the surface of the same pond. In short, as Davies puts it,[13]

(3.1) Organized waves get emitted, but only disorganized waves get absorbed.

The fact that emitters are associated with coherent radiation and absorbers apparently with incoherent radiation might suggest a different way of characterizing the asymmetry of radiation. That is, it might seem that the following is equivalent to 3·1:

(3·2) All emitters produce *retarded* rather than *advanced* wave fronts.

Many informal descriptions of the asymmetry of radiation suggest a formulation such as 3·2. Thus Davies again: "If I send a message through a radio transmitter to a distant colleague, or if I shout to him across the intervening space, I do not expect him to know my message *before* it is sent, but naturally presume that the radio or sound waves will travel outwards from the transmitter or my mouth, to reach a distant point at a later time."[14]

I don't mean to suggest that Davies and other writers who describe the asymmetry of radiation in such terms would deny that it is also correctly described by 3·1. The assumption seems to be that in 3·1 and 3·2 we have two different ways of describing the same state of affairs. There is an important difference between 3·1 and 3·2, however, which shows up when we ask what things would be like if radiation were *not* asymmetric. This is a crucial question, for if we want to explain the asymmetry of radiation it is vital to know what the "contrast class" is—in other words, to know what would count as not being in need of explanation. In the case of 3·1, our attention is directed to a stark difference between emitters and absorbers. Symmetry would therefore consist in the absence of this difference. There are actually several ways in which this might be achieved. Most obviously, it might be shown that contrary to appearances, absorbers are centered on coherent wave fronts after all. Call this

(3·3) Both emitters and absorbers are centered on coherent wave fronts (these being outgoing in the first case and incoming in the second).

There are a number of less obvious possibilities. For example, it might be shown that

(3·4) Neither emitters nor absorbers are centered on coherent wave fronts.

Alternatively (this possibility will play a role in a moment), it might be shown that although both emitters or absorbers are really centered on coherent wave fronts, these wave fronts are not simply retarded for emitters and advanced for absorbers, as we might expect: rather, in both cases they comprise some mixture of retarded and advanced wave fronts (the same mixture in both cases). In particular, it might be shown that

(3.5)   Emitters and absorbers are both centered on coherent wave fronts, *these being half outgoing and half incoming in both cases.*

All these possibilities would remove the apparent contrast between emitters and absorbers, and with it the temporal asymmetry of radiation, insofar as it is characterized by 3.1. Setting aside for the moment the alternatives 3.4 and 3.5, we may therefore say that according to formulation 3.1, the task of explaining the temporal asymmetry of radiation is best thought of as that of explaining why the following symmetric state of affairs does not obtain, or at least does not appear to obtain: that emitters and absorbers are *both* centered on coherent wave fronts (these being outgoing in the first case and incoming in the second).

However, let us now consider what symmetry would amount to when viewed in light of 3.2. If we read 3.2 as synonymous with 3.1, then of course it entails the same picture of symmetric radiation. But there is another possibility: 3.2 focuses our attention on emitters—i.e., on those physical events or entities that appear to be the source of coherent wave fronts—and notes that the associated wave fronts are always retarded rather than advanced. It is thus tempting to think that symmetry would consist in *those very entities* being centered equally on advanced and retarded wave fronts. (This might happen in one of two ways: in the statistical sense that in large ensembles of emitters, roughly half were advanced and half retarded; or in the individual sense that each emitter was associated equally with advanced and retarded wave fronts.) Thus 3.2 suggests the following description of the symmetric case (that is, of the state of affairs whose apparent absence in nature needs to be explained):

(3.6)   Emitters—those entities normally thought of as emitters—are associated equally with advanced and retarded wave fronts, in either the statistical or the individual sense.

Notice the contrast between this and 3.3. According to 3.3, symmetry requires no change in the nature (or apparent nature) of those actual physical systems we think of as emitters. It doesn't require that our radio transmitters "transmit into the past." It only requires that those quite different systems we think of as absorbers should *also* be associated with coherent radiation—i.e., in part, that our radio *receivers* should be centered on incoming coherent radiation.

An old-fashioned analogy might help at this point. Suppose we notice (or think we notice) that in a given community the men work and the women do not. We might describe the apparent gender imbalance by saying that in the community in question, all workers are male. However, this loose formulation allows two quite different ways of restoring symmetry. The first way, corresponding to 3·3, would be for the women to become workers (or to be recognized as workers already), on a par with the men. The second way, corresponding to 3·6, would be for half the existing male workers to become female—or for each existing male worker to become half male and half female!

Analogously, we have two quite different conceptions of what temporal symmetry would amount to for radiation. I don't know whether it is fair to say that physicists have had these confused, but it would not be surprising if they had done so, for the leading attempt to account for the apparent asymmetry of radiation does address itself to 3·6—at any rate at first sight. The central thesis of the Wheeler-Feynman Absorber Theory is that if we assume that the radiation associated with an isolated accelerated charged particle is *equally* advanced and retarded, we can explain why it *appears* to be fully retarded, in terms of the influence of distant material where the radiation is absorbed. The main thrust of the theory thus seems to accord with 3·6. We are told that the apparent asymmetry of radiation is manifest in the behavior of those events we call emitters, and that this behavior is *only* apparent—the underlying character of each of these events is symmetric with respect to time. (In other words, it is like being told that each of what we thought of as the male workers is "really" half male and half female.)

It is true that if the Wheeler-Feynman theory succeeds, then it shows not only what 3·6 requires, namely that emitters are intrinsically symmetric, but also that there is no intrinsic difference between so-called emitters and so-called absorbers. So it also establishes symmetry in the sense assumed by 3·1. It doesn't give us 3·3, but it does gives us 3·5. If successful, therefore, the Wheeler-Feynman theory does not disappoint those who correctly understand that 3·1 captures the mysterious asymmetry of radiation. At the same time, however, it offers a beguiling trap to anyone who misunderstands the mystery, taking it to consist in the absence of 3·6. For it panders to this misunderstanding, in giving us 3·6 as well as 3·5.

In my view, however, the Wheeler-Feynman theory is not successful as it stands. Two main flaws have been largely overlooked. One lies in a crucial argument concerning the link between the apparent asymmetry of radiation and that of thermodynamics. (It is our promised second instance of the double standard fallacy.) The other flaw concerns the status of one of the theory's implicit assumptions. I describe these difficulties below. However, my main aim is to show that the mathematical core of the Wheeler-Feynman theory can be reinterpreted so that it not only avoids these difficulties, but also provides a direct argument for 3·3. In my view this new interpretation encapsulates the real lesson of the Wheeler-Feynman argument—a lesson which seems to have been obscured, at least in part, by a tendency to construe the problem of the arrow of radiation in terms of 3·2 rather than 3·1.

THE WHEELER-FEYNMAN THEORY IN BRIEF

If we attempt to accelerate a charged particle, our efforts encounter resistance: we have to exert extra energy, compared to the case of an uncharged particle of the same mass. It is this extra energy which flows away into the electromagnetic field as light, radio waves, and other forms of electromagnetic radiation. The resistance to the original acceleration is called the radiative damping force, or simply the radiation resistance. Early attempts to understand this effect were hampered by mathematical difficulties associated with the idea that a charge could be affected by its own electric field.

In the early 1940s, Wheeler and Feynman tried to address these problems under the assumption that a charge does not interact with its own field. (Wheeler was then a young assistant professor at Princeton, and Feynman, barely younger, his graduate student.)[15] Their idea was that the radiative reaction might be due, not to the interaction of the charge with its own field, but to the effects of distant charges elsewhere. However, the progress which Wheeler and Feynman made in this original respect has been largely nullified by the apparent impossibility of extending its techniques successfully to the quantum mechanical case.[16] The main contemporary interest of their theory lies in its bearing on the issue of the arrow of radiation, and its connection with the arrow of thermodynamics. But the original motivation perhaps explains why Wheeler and Feynman made the otherwise unnecessary and inappropriate assumption that symmetry would require that individual accelerated charges radiate symmetrically into the past and the future—why they saw symmetry in terms of 3·6 rather than 3·3, in other words.

Motivation aside, the guiding idea of the Wheeler-Feynman argument is that even if radiation were actually symmetric in this sense—even if an accelerated charge did produce both an incoming and outgoing concentric wave front—the influence of this radiation on absorbing material elsewhere might be such as to give rise to the asymmetric result apparently observed. Their thought was that the radiation produced by the original acceleration would accelerate other charged particles when it encountered an absorber. These charged particles would radiate in turn, and the combined effect might be to produce the asymmetric result we actually observe.

With this end in mind, Wheeler and Feynman consider the effect on the charged particles of a future absorber of the full outgoing wave we normally expect to see from an accelerated charge i. This wave accelerates the charged particles of the absorber (see Figure 3.1), and Wheeler and Feynman assume

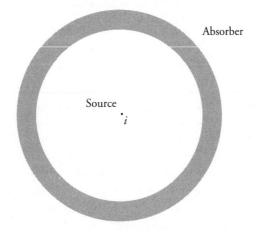

Figure 3.1. The setup for the Wheeler-Feynman argument.

that the secondary radiation produced by this acceleration is really one-half advanced, or incoming, and one-half retarded, or outgoing. (Remember that this is what the assumption that radiation is really symmetric amounts to, from their point of view.)

Wheeler and Feynman then "add up" the incoming waves associated with all the different particles of the absorber. Each of these individual waves is an incoming concentric ripple, centered on one of the particles in the absorber. All these waves occupy the space within the absorber at the same time, however, and Wheeler and Feynman ask what the combined effect amounts to. They show that in the space between i and the absorber, after the time of the original acceleration at i, the effect is to produce what looks exactly like an outgoing wave from the source, half the size of outgoing wave we normally expect to see.

Thus Wheeler and Feynman suggest that half of the outgoing wave we normally see can be attributed to the absorber response: it is "really" the combined advanced waves of the absorber particles. If the other half is what is "really" produced by the source i itself—as predicted by the assumption that the initial contribution from the source is also half retarded and half advanced—this seems to explain exactly what we normally observe. In other words, the result is interpreted as explaining the existence of a full retarded wave in the space between i and the absorber, after the time of the initial acceleration. This wave is thus attributed half to i itself and half to the response of the absorber.

In a similar way, Wheeler and Feynman propose that the observed absence of an incoming wave at i at the time of the initial acceleration is really the net result of destructive interference between a half-strength advanced wave from i and a half-strength advanced absorber response, whose phase is now such that the two waves exactly cancel out. The original acceleration at i really produces an incoming wave, on this view, but it is canceled by the combined advanced waves of the absorber particles, and so we don't see it.

In summary, then, Wheeler and Feynman suggest that the source contributes a radiation field of the form

1/2 retarded + 1/2 advanced,

the absorber contributes a field of the form

1/2 retarded − 1/2 advanced,

and the net result is therefore a retarded wave of the full value assumed originally. Accordingly, Wheeler and Feynman claim to have demonstrated that even if radiation is actually symmetric (again in the assumed sense), it nevertheless appears asymmetric, in virtue of the thermodynamic properties of the absorbers provided by the universe as a whole.[17]

WHY DOESN'T THE ARGUMENT WORK IN REVERSE?

Wheeler and Feynman note that one might employ the same chain of reasoning with the reverse temporal orientation. That is, one might try to show that the effect of a past absorber of the assumed half advanced wave from the source is to cancel the retarded wave from the source, and to build the advanced wave up to "full strength." If this worked then clearly the argument would be in trouble, for it would be claiming to show that the net wave from an accelerated charged particle in an absorbing enclosure is both fully retarded *and* fully advanced.

Wheeler and Feynman's answer is described concisely by Paul Davies:

> Absorption is clearly an irreversible thermodynamic damping effect; the entropy of the absorbing medium increases. This thermodynamic asymmetry in the absorber imposes an asymmetry on the electromagnetic radiation, by permitting the transport of energy from the source at the centre of the cavity to the cavity wall, but not the other way round. The advanced … solution, which is allowed on purely electrodynamic grounds, is thus ruled out as overwhelmingly improbable, because it would require the cooperative "anti-damping" of all the particles in the cavity wall … . Ions would become collisionally excited,

and radiate at the precise moment throughout the wall to produce a coherent converging wave to collapse onto the cavity centre at just the moment that the charged particle there was accelerated.[18]

Or as Wheeler and Feynman themselves put it, "Small *a priori* probability of the given initial conditions provides our only basis on which to exclude [the advanced case]."[19]

In effect, however, this is exactly the same argument that Popper used to attempt to explain why we don't see incoming water waves. Recall that his claim was that the conditions which would be needed around the edge of the pond are simply too unlikely. As we saw, however, this attempt to rely on a statistical explanation involves a temporal double standard. So long as we take into account statistical considerations alone, the probability of the arrangement of absorber particles required for incoming radiation is exactly the same as that required for outgoing radiation. So if a statistical argument rules out the advanced solution, an exactly parallel argument also rules out the usual retarded solution.

It might be objected that in the retarded case the probability is effectively determined not by statistical considerations but by a boundary condition: given that there is a retarded wave front from the source, in other words, there is in fact a very high probability that the absorber particles will come to be correlated in what would otherwise be an overwhelmingly unlikely fashion. This is true, but the basic assumption of the Wheeler-Feynman theory guarantees that same argument works in reverse! Recall that Wheeler and Feynman assume that in reality radiation is intrinsically "symmetric"—that is, half retarded and half advanced. So there *actually is* an advanced wave from the source, according to this view. (This is why this use of the double standard fallacy is even more blatant than the one we noted earlier in the chapter.) Given the existence of such a wave, it is no longer unlikely that there is an antidamping correlation of the particles in the cavity wall. On the contrary, and just as in the usual case, the boundary condition virtually guarantees such a correlation. So the statistical argument provides no grounds whatsoever for preferring the retarded self-consistent solution to the advanced one.

Note that it is no use objecting at this point that the advanced wave is eventually supposed to be canceled by the advanced wave from the future absorber. This is irrelevant here, for two reasons: first, because Wheeler and Feynman don't want to deny that the advanced wave exists, only to claim that because it is canceled we don't observe it; and second, more important, because to say that it is canceled is to assume the retarded solution whose priority is here in doubt. In plain terms, it is to beg the question.

As it stands, therefore, the Wheeler-Feynman argument fails to derive the apparent temporal asymmetry of radiation from that of thermodynamics. Indeed, if it were valid it would actually *disprove* the assumption of the underlying symmetry of radiation, for it would establish that this assumption leads to the contradictory conclusion that the net radiation inside an absorbing vessel surrounding an accelerated charged particle is both fully retarded *and* fully advanced!

ARE THE COMPONENTS DISTINCT?

The second problem I want to mention for the Wheeler-Feynman argument involves a physical assumption which Wheeler and Feynman require, but do nothing to justify—an assumption which when noticed seems ad hoc and implausible. The ability to avoid this assumption is another important advantage of the reinterpretation of the argument which I propose below.

Recall that the Wheeler-Feynman argument represents the original retarded wave between the source i and the absorber as a sum of two *equal* components, one identified as the one-half retarded wave from i and the other as the combined one-half advanced wave from the absorber. If we are to be justified in adding these components, however, we must have grounds for taking them to be *distinct* in the first place. Otherwise it is as if I were to deposit some money in my bank, withdraw it a month later, and claim interest both on the amount I had put in and on the amount I had taken out. My bank manager would politely point out that there were not two sums of money involved, but rather two different ways of describing the same single sum of money.

Similarly, if all we have in the Wheeler-Feynman case is two different ways of describing the same wave, then there are not really two components to constructively interfere with one another. Do Wheeler and Feynman have any justification for the claim that these component waves are actually distinct? It cannot lie in any physically measurable difference, for the waves concerned are qualitatively identical. Perhaps then the justification lies in the fact that the waves concerned have different sources. One component originates from i, and the other originates at the absorber (albeit under the influence of the retarded wave which itself originates at i). This suggestion lands the argument in further trouble, however. For in order to derive a response of the required magnitude from the absorber, the argument requires a full strength retarded wave from i. At this stage the *full* retarded wave needs to be treated as "sourced" at i. (After all, if we began with only the one-half retarded wave, the response of the absorber would also be halved—leaving us 25 percent short of our target, even if summing is allowed.) By the time the argument reaches its conclusion, one-half of this fully retarded wave is being accounted for as an advanced wave from the absorber. This in itself is not inconsistent, as long as we are prepared to allow that waves needn't have a unique source—that is, to allow that it is simply a matter of our own temporal perspective whether we say that the given wave originates at i or at the absorber particles. However, if we do allow that sources are nonunique and perspective-dependent in this way, we cannot then distinguish the one-half retarded wave from i and the one-half advanced wave from the absorber by appealing to a difference in their sources. In other words, we still have no reason to think that these two "components" are not simply one and the same wave, under two different descriptions. We still have no justification for "adding them up."

It is true that apparently ad hoc assumptions are often defended in science on the grounds that they give the right answers; if sufficiently successful, they come to seem less ad hoc. In the present case, however, we have seen that the original Wheeler-Feynman theory is already in trouble on other grounds. Moreover, as I now want to show, it is possible to interpret the Wheeler-Feynman argument so that it simply doesn't need the assumption that the two components are distinct.

THE NEW INTERPRETATION

Let us now consider the Wheeler-Feynman argument in light of propositions 3·1 and 3·3; that is, on the understanding that the real puzzle about radiation is that transmitters are centered on organized outgoing wave fronts, but receivers do not seem to be centered on organized incoming wave fronts. On this view the problem has nothing to do with the fact that transmitters are associated with outgoing rather than incoming radiation (which is no more puzzling than the fact that money deposited into a bank account turns up on one's balance after the transaction and not before). The puzzle simply concerns the fact that the radiation is organized in concentric wave fronts when it leaves a transmitter, but doesn't seem to be similarly organized when it arrives at a receiver.

Going back to Figure 3.1, let's begin in the same way as before, with the supposition that a transmitter at i radiates a fully retarded wave. Again we assume that i is surrounded by a shell of charged particles, and that these act as receivers, transferring energy from the field to nonelectromagnetic forms. Because they are receivers, we expect that from their point of view the radiation associated with this field is fully advanced, or incoming. However, let us now assume that *contrary to appearances,* this radiation is coherently centered on the absorber particles. In other words, we assume that each absorber particle is centered on what in the usual time sense looks like a converging coherent wave front.

The mathematical reasoning employed by Wheeler and Feynman then shows that in the region between i and the receiver, after the initial acceleration of i, this field is *equal in value* to the original wave from i. (The original argument had a factor of one-half at this point: the difference stems from the fact that the "response" of the receiver is now expected to be *fully* advanced.) How is this equality to be explained? By the fact, I suggest, that *these waves are one and the same*. (Here we are explicitly rejecting the ad hoc distinctness assumption which we objected to in the previous section.)[20]

In other words, I think the real lesson of the Wheeler-Feynman argument is that the same radiation field may be described equivalently either as a coherent wave front diverging from i, or as the sum of coherent wave fronts converging on the absorber particles. Interpreted in this way, the argument establishes 3.3: it shows that without inconsistency we may say that both transmitters and receivers are actually centered on coherent wave fronts. More generally, in the case of a free charged particle accelerated by incoming electromagnetic radiation, which then reradiates the received energy to the field, we may say that both the outgoing and the incoming radiation take the form of coherent wave fronts centered on the particle concerned.

WHY THE APPARENT ASYMMETRY?

According to this view the radiative asymmetry in the real world simply involves an imbalance between transmitters and receivers: large-scale sources of coherent radiation are common, but large receivers, or "sinks," of coherent radiation are unknown. Note that this is not to say that coherent sinks *as such* are unknown, or even uncommon, for on this view every individual absorber is such a sink. At the microscopic level things are symmetric, and we have both coherent sources and coherent sinks. At the macroscopic level we only notice the sources, however, because only they combine in an organized way in sufficiently large numbers.

In other words, it is rather as if a bank account were to gain much of its funds from a relatively small number of very large deposits, but to lose them only to a very large number of very small withdrawals. There would certainly be a temporal imbalance in such a case, but it wouldn't lie in the banking processes themselves. The microtransactions would be symmetric, microwithdrawals being the temporal inverse of microdeposits. The macroscopic difference would arise because microdeposits clumped together in an orderly way to form macrotransactions, whereas microwithdrawals did not (or not to such extent). To explain the macroasymmetry we would need to look at the bank's connections to the outside world—at why there was large-scale organization of deposits, but no corresponding organization of withdrawals.

Similarly in the case of radiation. What needs to be explained is why there are large coherent sources—processes in which huge numbers of tiny transmitters all act in unison, in effect—and why there are not large coherent absorbers or sinks, at least in our region of the universe. Of course, by now we know that the former question is the more puzzling one, for it is the abundance of large sources rather than the lack of large sinks which is statistically improbable, and associated with the fact that the universe is very far from thermodynamic equilibrium. In a universe in thermal equilibrium, large sources and large sinks would be equally uncommon. All transmitters and receivers would be microscopic and uncorrelated with one another, except by pure chance. The thermodynamic field would be something like the surface of a pond in a heavy rainstorm: an incoherent mess of tiny interactions, with no noticeable pattern, in either temporal direction. In order to explain why our universe is not like this, what needs to be explained is the existence of such things as stars and galaxies, for they provide the electromagnetic equivalent of giant rocks (or otters!), which impose a pattern on the underlying disorder.

As I promised, then, this reinterpretation of the Wheeler-Feynman argument leads to the conclusion that the asymmetry of radiation has the same cosmological origins as the thermodynamic asymmetry. Unlike the standard Wheeler-Feynman argument, however, the route to this conclusion does not involve an attempt to explain the radiative asymmetry *in terms of* the thermodynamic asymmetry. In particular, the present argument does not depend on the thermodynamic properties of the absorber. (As we are about to see, it does not depend on the presence of the absorber at all.) The radiative arrow becomes not so much a child of the second law as a junior sibling. The arrow of radiation guides us not *to* the arrow of thermodynamics, but in the same direction. Both lead us to the same kind of question about our distant past: Why does the universe have this kind of history—a history that produces stars and galaxies, for example—and not a more likely one? And why does it do so in such an asymmetric way? (For example, why don't we find "sink stars," on which radiation converges, as well as the usual kind.)

We'll return to these cosmological issues in the next chapter. Before leaving the Wheeler-Feynman theory, however, I want to note another connection between the arrow of radiation and cosmology—or rather, the absence of a connection, which seems to be a further advantage of my proposed reconstruction of the Wheeler-Feynman argument. I also want to comment briefly on the impact of my approach on some spin-offs of the Wheeler-Feynman argument in physics. But the next two sections are mainly for specialists. Other readers may skip to the summary at the end of the chapter.

NO NEED FOR A FUTURE ABSORBER

We have seen that in the original Wheeler-Feynman argument, the observed outgoing wave is thought of as the sum of two equal components, one from the source and the other from the absorber. So the presence of the absorber is crucial. Without it, the same reasoning would suggest that we would observe only a half-strength outgoing wave from the source—and also a half-strength incoming field, since this would now not be canceled by the advanced field from an absorber. It is true that Wheeler and Feynman suggest a way to avoid this consequence. This involves other problems, however, for it requires the presence of explicit advanced effects.

Thus the orthodox version of the Absorber Theory gives us a choice between, on the one hand, a very strong cosmological constraint, namely that the future universe is an (almost) perfect absorber, and on the other hand, the conclusion that insofar as the future universe is transparent, a corresponding amount of source-free radiation needs to be postulated in the early universe. It is true that some physicists have seen this as an argument for particular cosmological theories.[21] However, most have taken the view that it is a disadvantage of the theory that it imposes such a close connection between electromagnetic radiation and cosmology.

Thus it is a further attraction of my proposal that it doesn't seem to have any such consequence. It allows us to say that where the absorber is absent, the retarded wave from the source particle *i* simply propagates to future infinity. What matters is that to whatever extent there is an absorber, its interaction with the field may be consistently redescribed in terms of coherent waves converging on the absorber particles. The general result will be that the combined contribution of these waves, when taken together with the remaining unabsorbed component, yields the same physical entity as does the original description. Thus the crucial difference is that the usual retarded wave is no longer taken to *need* two (finite) sources, one in the past and one in the future; the claim is simply that insofar as such a wave does have two such sources, their contributions are entirely consistent. Time-symmetric electromagnetism is freed from the constraints of cosmology.

RELATED ISSUES IN PHYSICS

The classical Absorber Theory has been extended and applied in novel ways by a number of authors. In this section I want to make some brief comments on the likely bearing of the above argument on some of these extensions and applications.

The basis of the proposed reinterpretation of the Wheeler-Feynman argument is the recognition that radiative symmetry does not require that radiative emitters be *individually* symmetric in time, in the sense of 3·6. Symmetry would also be secured if the class of emitters of retarded radiation turned out to be "mirrored" by a class of absorbers of advanced radiation. As reinterpreted, the Wheeler-Feynman argument shows that this latter kind of symmetry is mathematically consistent: no inconsistency is generated by taking the advanced solutions of the electromagnetic wave equations to characterize absorbers (which, like emitters, are thus represented as being centered on coherent wave fronts).

Hence there need be nothing asymmetric in taking the radiation associated with emitters to be fully retarded; and insofar as it seems to deny this, the Wheeler-Feynman Absorber Theory involves a misleading conception of radiative asymmetry. Some proposed extensions of the Absorber Theory appear to be guided by this conception. This seems particularly true of various attempts to generalize the Absorber Theory to an account of the interactions of particles other than photons.[22] These approaches are guided by the Absorber Theory's 3·6-based conception of temporal symmetry, and hence by the belief that there is something asymmetric in standard models of particle emission. As Paul Csonka puts it, "We wish to consider all particles to be on an equal footing ... and to construct a theory in which all are emitted, and absorbed, in a time-symmetric manner."[23] In light of the above reassessment of the Absorber Theory, however, it is difficult to avoid the conclusion that these proposals address a nonexistent problem. Within well-known limits (involving the behavior of neutral kaons), particle physics is temporally symmetric as it stands. Absorption is the temporal inverse of emission, and symmetry does not require that the two kinds of event be rolled into one.

The view that standard particle physics is temporally asymmetric has perhaps also been encouraged by Feynman's rather loose talk of particles "traveling" backwards and forwards in time. If this idea is taken literally then it permits a distinction between the emission of a particle going "backwards in time" and the absorption of a particle "going forwards in time." From this it is a short step to the view that conventional particle physics describes the latter kind of event but not the former. In my view, however, the notion that particles are intrinsically directed in time is as problematic as the corresponding doctrine about radiation. Both particles and waves are simply retarded (outgoing) with respect to their point of emission; and advanced (incoming) with respect to their point of absorption. This is not a physical fact, but just a consequence of the meanings of the terms involved. It remains true if the temporal framework is reversed, though in this case the labels are interchanged: emissions are construed as absorptions, incomings as outgoings, and so on.[24]

Finally, I want to comment briefly on a related but quite differently motivated generalization of the Absorber Theory to particle interactions and quantum mechanics. John Cramer, of the University of Washington, has suggested in recent years that we may apply a notion analogous to that of the absorber response in order to account for EPR phenomena and some of the other "paradoxes" of quantum mechanics.[25] The crucial feature of Cramer's "transactional" interpretation of quantum mechanics is that the physical state of a system between an emitter and a measuring apparatus depends not only on its standard quantum mechanical state vector (considered as a real physical wave), but also on an advanced "response wave" which the standard retarded wave generates from the future measuring apparatus. The state at intervening times is a product of both these factors. (Cramer offers the analogy of a standing wave generated in a cavity as an initial wave interacts with its reflections.)

As Cramer notes, his theory belongs to a family of interpretations based on the admission of "advanced action" or "backward causation" of some kind. Cramer is right, in my view, in thinking that these approaches are very attractive, and unjustly neglected by physicists and philosophers. Indeed, this is one of the main themes of the second part of this book: I think that the single most important consequence of a proper appreciation of an atemporal Archimedean standpoint in physics is to give this approach the visibility it deserves. At present, Cramer's theory is probably the most highly developed version of such an approach to quantum mechanics. It would be unfortunate if such a well-motivated theory turned out to depend on what otherwise seems an unnecessary and misleading account of temporal symmetry.

However, it is not clear that Cramer's proposal does depend in any essential way on the Wheeler-Feynman conception of temporal symmetry. As just noted, the crucial assumption is that the state of a system at a time may depend on its future as well as on its past. But in a sense this is also a feature of the reinterpreted Wheeler-Feynman argument: in its new form the argument shows that an electromagnetic radiation field may be taken to be determined either by its past sources or by its future sinks (absorbers); the two representations give equivalent results. Cramer's theory seems to be amenable to reinterpretation in a similar way, a move that would parallel the reinterpretation of the classical Absorber Theory proposed in this chapter. As I'll explain in chapters 8 and 9, there are other respects in which Cramer's approach seems to miss the most appealing route to an advanced action view of quantum mechanics. Insofar as it builds on the classical Absorber Theory, however, I think it is comparatively untroubled by the arguments of this chapter.

SUMMARY

1. It is important to avoid some confusions about what the asymmetry of radiation actually amounts to. The puzzle doesn't lie in the fact that radio transmitters don't transmit into the past, for example. It lies in the fact that the waves leaving transmitters are organized, while the waves arriving at receivers seem to be disorganized.

2. Once the nature of the asymmetry is properly understood, the Wheeler-Feynman argument can be taken to show that what we observe is quite compatible with the hypothesis that radiation is intrinsically symmetric at the micro level: that microemitters and microabsorbers are both centered on organized wave fronts. The apparent asymmetry arises because microemitters "clump together" coherently into a macroscopic sources, in a way in which macroabsorbers do not. Interpreted in this way, the Wheeler-Feynman argument escapes two major objections which may be raised against the standard version of the argument.

3. This diagnosis of the observed asymmetry of radiation shows that what needs primarily to be explained is the clumping of emitters into macroscopic sources, not the lack of clumping of sinks. In a statistically "natural" world, we wouldn't find clumping of either sort. Again, therefore, the main explanatory focus is on the past rather than the future. In particular, we must resist the temptation to think (a) that what needs to be explained is why we don't find incoming coherent radiation (in the usual time sense), and (b) that the explanation lies in statistical arguments. For step (b) involves the double standard fallacy, and step (a) ignores the real puzzle, which is why we do see incoming coherent radiation if we look in the opposite temporal direction. (The time to think about [a] is when we come to try to explain the ordered past in cosmological terms. At that stage the question whether the future might be similarly ordered—and if not, why not—becomes an important one, as we shall see in chapter 4.)

4. The arrow of radiation thus provides a further important case of the general tendency displayed by the arrow of thermodynamics: the tendency of the physical world to display a very remarkable degree of coordinated behavior in the temporal direction we think of as future-to-past. Once again, it seems that we must look to cosmology for an explanation of this behavior. We want

to know why the universe contains the systems on which this coordinated behavior is centered—what we ordinarily think of as the *sources* of coherent radiation—including such things as stars and galaxies.

5. Finally, a more general point that emerges from this discussion. The model of macroscopic asymmetry built on microscopic symmetry provides an important insight into some issues which will be important later in the book. Recall PI[3], the Principle of Independence of Incoming Influences, which we encountered in chapter 2. The asymmetry of radiation has often been taken to provide a particularly clear illustration of the operation of this principle. Intuitively, an incoming concentric wave would seem to involve a violation of PI[3], because the distant segments of such a wave front are highly correlated one with another. Similarly, ordinary retarded waves have been taken to provide a natural illustration that outgoing influences are typically not independent. In the case of radiation, however, it turns out that the familiar macroscopic asymmetry is compatible with microscopic symmetry. This suggests that PI[3] itself might be valid only at the macroscopic level. As we shall see later on, this provides an important clue to the relationship between PI[3] and our intuitive picture of causality in the world, and perhaps the missing secret of quantum mechanics itself.

4

←——————→

Arrows and Errors in Contemporary Cosmology

A CENTURY or so ago, Ludwig Boltzmann and other physicists attempted to explain the temporal asymmetry of the second law of thermodynamics. As we saw in chapter 2, the hard-won lesson of that endeavor—a lesson still commonly misunderstood—was that the real puzzle of thermodynamics is not why entropy increases with time, but why it was ever so low in the first place. To the extent that Boltzmann himself appreciated that this was the real issue, the best suggestion he had to offer was that the world as we know it is simply a product of a chance fluctuation into a state of very low entropy. We saw that his statistical treatment of thermodynamics implied that although such states are extremely improbable, they are bound to occur occasionally, if the universe lasts a sufficiently long time. Ingenious as it is, we saw that this is a rather desperate solution to the problem of temporal asymmetry. Not least of its problems is the fact that it implies that all our historical evidence is almost certainly misleading. This is because the "cheapest" or most probable fluctuation compatible with our present experience is one which simply creates the world as we find it, rather than having it evolve from an earlier state of even lower entropy.

One of the great achievements of modern cosmology has been to offer an alternative to Boltzmann's desperate solution. It now seems clear that temporal asymmetry is cosmological in origin, a consequence of the fact that entropy was extremely low soon after the big bang. The universe seems to have been in a very special condition at this stage in its history—so special, in fact, that some ten to twenty billion years later, the effects are still working their way through the system, giving us the conditions required for the various familiar asymmetries which we have been discussing in the last two chapters. To a large extent, then, Boltzmann's problem seems to reduce to that of explaining this special condition of the early universe.

Ironically, however, when contemporary cosmologists think about why the universe started in this special way, they often make the same kind of mistakes as were made in Boltzmann's time. In this chapter I want to show that these mistakes are widespread in modern cosmology, even in the work of some of the contemporary physicists who have been most concerned with the problem of the cosmological basis of the arrow of time. (We'll see that even writers as distinguished as Stephen Hawking, Roger Penrose, and Paul Davies do not seem to be immune.) Once again, the secret to avoiding these mistakes is to learn to approach the issue from an atemporal Archimedean standpoint. It turns out, I think, that cosmology is farther away from an adequate explanation of temporal asymmetry than many cosmologists believe. However, the Archimedean standpoint at least gives us a clear view of what still needs to be done.

THE NEED FOR SMOOTHNESS

What exactly is special about the early universe? Mainly the fact that the matter it contained seems to have been distributed in an extremely smooth and homogenous way, with almost the same density everywhere. At first sight, it seems surprising to call this a special or unusual state of affairs. Isn't it simply the most natural or disordered possibility? After all, we expect a gas to spread itself uniformly throughout its container. But this is because the dominant force in the case of a gas is the pressure its molecules exert on one another, which tends to push them apart. In the case of the universe as a whole, however, the dominant force is gravity, which tends to pull matter together. The "natural" state for a system dominated by gravity is thus a clumpy one, in which the gravitational force has caused the material in question to collect together in lumps. (Another case in which the dominant force is attractive is the behavior of a film of water on a flat waxy surface. Here the main force is surface tension, and the normal behavior, as we all know, is for the water to collect in droplets, rather than spreading uniformly over the surface.)

Why is a smooth distribution of matter in the early universe so important? Because without it, galaxies would not have been able to form. Galaxies are very small clumps of matter, on the scale of the universe as a whole. If the early distribution of matter had not been so smooth, gravity would have tended to collect matter into vastly larger clumps (huge black holes, in fact). So the kind of universe we presently observe would simply not have been possible. Smoothness is very important, then, though it turns out it is possible to overdo things: if the universe had been too smooth, even relatively small clumps of the size of galaxies would not have formed.[1]

Initially, the conclusion that the early universe must have been (extremely, but not completely) smooth was derived from theoretical models of the evolution of the universe: in terms of these models, smoothness turned out to be a necessary precondition for the kind of universe we presently observe. Very recently, however, the conclusion has received some dramatic observational support. Since the 1960s, astronomers have known of the microwave background radiation, itself a remnant of processes in the early universe. Satellite-based studies of this radiation have now confirmed not only that this radiation is very similar from all directions in space, as smoothness predicts; but also that there are indeed tiny variations in different directions, of just the level predicted by the minute inhomogeneities required for galaxy formation.[2]

In recent years, then, cosmologists have been trying to explain why the matter in the early universe was arranged in this smooth way. There are two aspects to this problem: Why is the early universe so smooth? And why are there the minor irregularities required for galaxy formation? However, the first of these issues is the more directly connected to problems of temporal asymmetry. In particular, as we shall see, it is the smoothness of the early universe which looks puzzling when we look at these issues from an atemporal standpoint. In effect, the smoothness of the early universe is the big puzzle of temporal asymmetry, to which the little puzzles of the previous chapters all seem to lead. Smoothness alone does not account for all features of the familiar temporal asymmetries, but it seems to be an essential and very powerful prerequisite for all of them.

In this chapter I want to look at some recent attempts to explain the universe's early smoothness. It turns out that cosmologists who discuss these issues often make mistakes very similar to those we have encountered in the two previous chapters. In particular, they tend to fall for the temporal double standard: the mistake of failing to see that certain crucial arguments are blind to temporal direction, so that any conclusion they yield with respect to one temporal direction must apply with equal force with respect to the other. We saw that writers on thermodynamics often failed to notice that the statistical arguments they sought to employ are inherently insensitive to temporal direction, and were hence unable to account for temporal asymmetry. And writers who did notice this mistake commonly fell for another: recognizing the need to justify the double standard—the application of the arguments in question "toward the future" but not "toward the past"—they appealed to additional premises, without noticing that in order to do the job, these additions must effectively embody the very temporal asymmetry which was problematic in the first place. To assume the uncorrelated nature of initial particle motions or incoming "external influences," for example—our principle PI[3]—is simply to move the problem from one place to another. (It may *look* less mysterious as a result, but this is no real indication of progress. One of the fundamental lesson of these attempts is that much of what needs to be explained about temporal asymmetry is so commonplace as to go almost unnoticed. In this area more than most, ordinary intuitions are a very poor guide to explanatory priority.)

In this chapter I want to show that mistakes of these kinds are widespread in modern cosmology, even in the work of some of the contemporary physicists who have been most concerned with the problem of the cosmological basis of the arrow of time. Interdisciplinary point-scoring is not the main aim, of course. By drawing attention to these mistakes I hope to clarify what would count as an adequate cosmological explanation of temporal asymmetry. And as in the previous chapters, I want to recommend the adoption of the atemporal Archimedean standpoint as an antidote to temporal double standards.

GOLD UNIVERSES AND THE BASIC DILEMMA

It is well known that if the universe is sufficiently dense, the force of gravity will be sufficient to overcome the present expansion. In this case the universe will eventually begin to contract again, and accelerate toward a final cataclysmic endpoint, the so-called big crunch. What would such a big crunch be like? Would it be a kind of mirror image of the big bang, at least in the sense that it too would display the same remarkable smoothness? On the face of it, it is hard to see how time-symmetric physical theories could imply that one end of the universe has to be "special" in this way, without implying that both ends have to be special.

The suggestion that entropy might be low at both ends of the universe was made by the cosmologist Thomas Gold in the early 1960s.[3] Gold's hypothesis was prompted by the appealing idea that there might be a deep connection between the expansion of the universe and the second law of thermodynamics. An early suggestion was that the expansion itself might increase the maximum possible entropy of the universe, in effect by creating new possibilities for matter. The thermodynamic arrow might simply be the result of the process in which the universe takes up these new possibilities. Gold saw that this would imply that in a contracting universe, the reverse would happen: entropy would decrease, because the contraction reduces the total stock of possibilities.

The idea that the arrow of thermodynamics is linked in this directed way to the expansion of the universe turns out to be untenable, however. The main objection to it is that the smooth early universe turns out to have been incredibly "special," even by the standards prevailing at the time. Its low entropy doesn't depend on the fact that there were fewer possibilities available, in other words. All the same, Gold's symmetric model of the universe has a certain appeal.

On the whole, however, cosmologists have not taken Gold's hypothesis very seriously. Most think it leads to absurdities or inconsistencies of some kind. But it turns out these objections to Gold's view often rely on a temporal double standard. People argue that if Gold were right, matter would have to behave in extremely unlikely ways as entropy deceased. They fail to appreciate that what Gold's view requires toward the future is just what the standard view requires toward the past. We know that the argument doesn't work toward the past: entropy does decrease in that direction, despite the fact that its doing so requires that matter be arranged in what seems an extremely unlikely fashion. So we have no reason to think that exactly the same argument is conclusive toward the future, in ruling out Gold's symmetric universe.

Because cosmologists often fail to notice double standards of this kind, they often fail to appreciate how little scope there is for an explanation of the smoothness of the big bang which does not commit us to the Gold universe. On the face of it, as I noted earlier, a symmetric physics seems bound to lead to the conclusion either that both ends must be smooth (giving the Gold universe), or that neither end need be, in which case the smooth big bang remains unexplained. On the face of it, then, we seem to be presented with a choice between Gold's view, on the one hand, and the conclusion that the smooth big bang is inexplicable (at least by a time-symmetric physics), on the other. Most cosmologists regard both options as rather unsatisfactory, but I think they fail to appreciate how difficult it is to steer a course between the two.

I call this choice the *basic dilemma,* and I want to show how double standards have tended to hide it from view. I shall also discuss some of the ways in which cosmology might be able to avoid the dilemma—to steer a middle course, in effect. And finally, I want to take another look at Gold's view. I'll give some examples to show how the standard objections rest on temporal double standards; and I'll briefly discuss some of the fascinating issues that arise if we take the view seriously. Could we already observe the kind of time-reversing future that Gold's view predicts, for example?

SMOOTHNESS: HOW SURPRISING IS IT?

The modern cosmological descendant of the puzzle Boltzmann was left with—Why was entropy low in the past?—is the question why the universe is so smooth in its early stages, after the big bang. To ask this question is to call for an explanation of an observed feature of the physical universe, in effect. A natural response to such a challenge is to try to show that the feature in question isn't surprising or improbable—that things "had to be like that." Accordingly, at least in the early days of modern cosmology, many cosmologists were inclined to argue that the state of the early universe is not really particularly special. For example, the following remarks are from one of Paul Davies' early accounts of cosmology and time asymmetry:

> It is clear that a time-asymmetric universe does not demand any very special initial conditions. It seems to imply a creation which is of a very general and random character at the microscopic level. This initial randomness is precisely what one would expect to emerge from a singularity which is completely unpredictable.[4]

The writer who has done the most to draw attention to the error in this general viewpoint is the Oxford mathematician Roger Penrose. Penrose's strategy is to ask what proportion of possible states of the early universe exhibit the degree of smoothness apparent in the actual early universe. He approaches the issue by comparing estimates of the entropy associated with a smooth arrangement of matter to that associated with alternative possible arrangements, in which most of the matter is concentrated in large black holes. Drawing on the fact that entropy is related to probability, he arrives at a figure of 1 in $10^{10^{123}}$—in other words, at the conclusion that in purely statistical terms, the actual early universe was special, or unnatural, to this stupendous degree.[5]

Penrose's argument relies on the fact that the entropy of a black hole is vastly greater than that of a spatially homogeneous arrangement of the same amount of matter. This fact in turn relies on some factors I mentioned earlier. At cosmological scales, the main force acting on the material in the universe is that of gravity. Because gravity is an attractive force, its "natural" or equilibrium state is one in which matter is drawn together as completely as possible—in contrast, say, to the case of a gas, where the repulsive effects of pressure encourage the gas to become more dispersed with time.

As we saw, these facts about gravity provide a slightly different way to make the point that the smooth early universe is statistically exceptional—an argument more closely related to the central themes of this book. It is to note that we would not regard a collapse to a smooth *late* universe, just before a big crunch, as statistically natural—quite the contrary, in fact. In the absence of any prior reason for thinking otherwise, however, this consideration applies just as much to one end of the universe as to the other. In these statistical terms, then, with double standards disallowed, a smooth big bang is just as unlikely as a smooth big crunch.[6] And how could it be otherwise? After all, until we find some reason to think otherwise, we should take the view that it is not an objective matter which end of the universe is the "bang" and which the "crunch."

In view of the importance of this point it is worth spelling it out in a little more detail, and giving it a name. I'll call it the *gravitational symmetry argument*. The argument has three main steps:

1. We consider the natural condition of a universe at the end of a process of gravitational collapse. In other words, we ask what the universe might be expected to be like in its late stages, when it collapses under its own weight. As noted above, the answer is that it is likely to be in a very inhomogeneous state, clumpy, rather than smooth.

2. We reflect on the fact that if we view the history of our universe in reverse, what we see is a universe collapsing under its own gravity, accelerating toward a final crunch.

3. We note that there is no objective sense in which this reverse way of viewing the universe is any less valid than the usual way of viewing it. Nothing in physics tells us that there is a wrong or a right way to choose the orientation of the temporal coordinates. *Nothing in physics tells us that one end of the universe is objectively the start and the other end objectively the finish.* In other words, the perspective adopted at (2) is just as valid for determining the natural condition of what we usually call the early universe, as the standard perspective is for determining the likely condition of what we call the late universe.

The main lesson to be learned from this argument is that there is much less scope for differentiating the early and late stages of a universe than tends to be assumed. If we want to treat the late stages in terms of a theory of gravitational collapse, we should be prepared to treat the early stages in the same way. Or in other words, if we do treat the early stages in some other way—in terms of some additional boundary constraint, for example—then we should allow that the late stages may be subject to the same constraint. On the one hand, this means that we cannot take for granted that the smooth early universe is unexceptional; on the other, it means that we cannot conclude that a smooth late universe (as Gold's view requires) is impossible.

This point is easy to appreciate, I think, once we learn to approach the problem of temporal asymmetry in cosmology from an atemporal point of view—in particular, once we learn to challenge the idea that there is something privileged about our ordinary temporal perspective. By and large, however, contemporary cosmology has not managed to make this conceptual leap, or at least has not managed to apply it consistently.

In support of this claim, I want to discuss two important recent suggestions about the origins of cosmological time asymmetry. In different ways, I think, both examples reflect a failure to address the issue from an appropriately atemporal stance.

THE APPEAL TO INFLATION

The first case stems from what cosmologists call the *inflationary model*. This theory is best thought of as a kind of "front end" to the standard big bang model, describing what might have happened to the universe in its extremely early stages. The basic idea is that when the universe is extremely small, and extremely young—and perhaps simply the product of some chance quantum fluctuation—the physical forces it experiences are very different from those with which we are familiar. In particular, the force of gravity is repulsive under these conditions, rather than attractive, and the effect is that the universe undergoes a period of exponential expansion: over and over again, in time intervals of the order of 10^{-34} seconds, the universe doubles its size. As it grows it cools, however, and at a certain point it undergoes what is called a *phase transition*. At this point a transformation takes place, which is somewhat analogous to that of steam into water, or water into ice. The fundamental forces change, gravity becomes attractive, and the universe settles into the more sedate expansion of the "ordinary" big bang.[7]

Since it was first proposed in the early 1980s, one of the main attractions of this inflationary model has been that it seems to explain a range of features of the early universe which the standard big bang model simply has to take for granted. One of these features, it is claimed, is the smoothness of the universe after the big bang. However, the argument that inflation explains smoothness is essentially a statistical one. The crucial idea is that during the inflationary phase the repulsive gravity will tend to "iron out" inhomogeneities, leaving a smooth universe at the time of the transition to the classical big bang. In 1983 this argument was presented by Paul Davies in an article in *Nature*. Davies concludes that "the Universe ... began in an arbitrary, rather than remarkably specific, state. This is precisely what one would expect if the Universe is to be explained as a spontaneous random quantum fluctuation from nothing."[8]

This argument illustrates the temporal double standard that commonly applies in discussions of these problems. To call attention to the fallacy, we need only note that as in step (2) of the gravitational symmetry argument, we might equally well view the problem in reverse. From this perspective we see a gravitational collapse toward what we ordinarily call the big bang. In statistical terms, this collapse may be expected to produce *inhomogeneities* at the time of any transition to an inflationary phase (which will now appear as a deflationary phase, of course). *Unless one temporal direction is already privileged, the statistical reasoning involved is as good in one direction as the other.* Hence in the absence of a justification for the double standard—a reason to apply the statistical argument in one direction rather than the other—the appeal to inflation cannot possibly do the work required of it.[9]

Davies also argues that

> a recontracting Universe arriving at the big crunch would not undergo "defla-
> tion," for this would require an exceedingly improbable conspiracy of quan-
> tum coherence to reverse-tunnel through the phase transition. There is thus
> a distinct and fundamental asymmetry between the beginning and the end of
> a recontracting Universe.[10]

However, he fails to notice that if double standards are kept at bay, this point conflicts with the argument he has given us concerning the other end of the universe: viewed in reverse, the transition from the ordinary big bang to the inflationary phase involves exactly this kind of "improbable conspiracy." If deflation is unlikely at one end, then inflation is unlikely at the other. Again, this follows immediately from the realization that there is nothing objective about the temporal orientation. A universe that collapses without deflation just *is* a universe that expands without inflation. It is exactly the same universe, under a different but equally valid description.[11] So the atemporal standpoint reveals that Davies' argument cannot possibly produce the kind of asymmetric conclusion he claims to derive.

HAWKING AND THE BIG CRUNCH

The second case I want to discuss is better known, having been described in Stephen Hawking's best-seller *A Brief History of Time.*[12] It is Hawking's own proposal to account for temporal asymmetry in terms of what he calls the *no boundary condition,* which is a proposal concerning the description of the universe in quantum mechanical terms. To see what is puzzling about Hawking's views, let us keep in mind what I called the basic dilemma: provided we are careful to avoid double standard fallacies, it seems on the face of it that any argument for the smoothness of the universe will apply at both ends of a recollapsing universe or at neither. So our choices seem to be to accept the globally symmetric Gold universe, or to resign ourselves to the fact that temporal asymmetry is not explicable by a time-symmetric physics, at least without additional assumptions or boundary conditions invoked for the purpose. The dilemma is particularly acute for Hawking, because he has more reason than most to avoid resorting to additional boundary conditions. They conflict with the spirit of his no boundary condition, namely that one restrict possible histories for the universe to those that "are finite in extent but have no boundaries, edges, or singularities."[13]

Hawking is one of the few contemporary cosmologists actually to have endorsed Gold's proposal, at least temporarily. In *A Brief History of Time* he tells us that he thought initially that the no boundary condition favored Gold's view: "I thought at first that the no boundary condition did indeed imply that disorder would decrease in the contracting phase."[14] He changed his mind, however, in response to objections from Don Page and Raymond Laflamme. As Hawking says,

> I realized that I had made a mistake: the no boundary condition implied that disorder would in fact continue to increase during the contraction. The thermodynamic and psychological arrows of time would not reverse when the universe begins to contract or inside black holes.[15]

By changing his mind in this way, Hawking avoids the objections which have led most other cosmologists to reject Gold's view. In their place, however, he runs into the problems associated with the second horn of the basic dilemma. In other words, he needs to explain how he has managed to get an asymmetric conclusion from what professes to be a time-symmetric physical theory. He needs to explain how his proposal can imply that entropy is low near the big bang, without equally implying that it is low near the big crunch. Hawking tells us that he has managed to do this, but doesn't explain how.

This puzzled me when I first read *A Brief History of Time* in 1988. I wrote to Hawking to ask about it, and sent copies of the letter to Page and Laflamme. None of them replied to this skeptical philosopher from the other side of the world, and so I wrote a short article explaining my puzzlement. I sent it off—very optimistically, I thought at the time—to the journal *Nature,* which published it in July 1989.[16] I was heartened to receive some supportive letters from physicists, including one from Dieter Zeh, himself the author of an important book on the physics of time asymmetry which appeared in 1989,[17] who wrote to say that he agreed with every word of my article. There was still no direct response from Hawking's camp, but my article was mentioned in a news column in *Scientific American* the following October, with some comments from Hawking and Page. As far as I could see, however, these comments did nothing to clear up the mystery as to how Hawking had managed to achieve what he claimed to have achieved.

Since then, however, a consensus seems to have emerged as to what Hawking's solution is, if he has one. A number of writers have pointed out that there is an important loophole which allows a symmetric physical theory to have asymmetric consequences, and suggested that Hawking's argument takes advantage of this loophole. The writers are quite right about the loophole: it rests on a very simple point, as I'll explain in a moment. But I think they are wrong to suggest that it removes the puzzle from Hawking's account. As I want to show, the real mystery is how, if at all, Hawking manages to avail himself of this loophole.

First, then, to the nature of loophole. The easiest way to get an idea of what Hawking has to have established is to think of three classes of possible universes: those which are smooth and ordered at both temporal extremities, those which are ordered at one extremity but disordered at the other, and those which are disordered at both extremities. Let us call these three cases *order-order* universes, *order-disorder* universes, and *disorder-disorder* universes, respectively. (Keep in mind that in the absence of any objective temporal direction we could just as well call the second class the disorder-order case.) If Hawking is right, then he has found a way to exclude disorder-disorder universes, without thereby excluding order-disorder universes. In other words, he has found a way to ensure that there is order at at least one temporal extremity of the universe, without thereby ensuring that there is order at both extremities. Why is this combination the important one? Because if we cannot rule out disorder-disorder universes then we haven't explained why our universe is not of that sort; while if we rule out disordered extremities altogether, we are left with the conclusion that Hawking abandoned, namely that order will increase when the universe contracts.

Has Hawking shown that order-disorder universes are overwhelmingly the most probable? It is important to appreciate that this would not be incompatible with the underlying temporal symmetry of the physical theories concerned. This is the loophole: a symmetric physical theory might be such that all or most of its possible realizations were individually asymmetric. The point can be illustrated with some very familiar analogies. Think of a factory which produces equal numbers of left-handed and right-handed corkscrews, for example. Each individual corkscrew is spatially asymmetric, but the production as a whole is completely unbiased. Or think of an organization whose employment practices show no bias at all between men and women: the policy as a whole is unbiased, but each individual employee is either male or female. In principle the same kind of thing might be true with respect to temporal asymmetry: a time-symmetric physical theory might have the consequence that any individual universe has to be asymmetric in time.

This possibility represents a very important loophole in the basic dilemma, then. In principle, as I already recognized in my *Nature* article, Hawking might have succeeded in exploiting it. In other words, he might have shown that the no boundary condition implies that all (or most) possible histories for the universe are asymmetric, with low entropy (order) at one end, and high entropy (disorder) at the other. A recent paper by J. J. Halliwell, one of Hawking's early collaborators, describes Hawking's view in just these terms; and in his recent book *About Time,* Paul Davies mentions my criticism and says that this loophole provides Hawking's way out.[18]

If this is Hawking's conclusion, however, he hasn't yet explained to his lay readers how he manages to reach it. It seems clear that it cannot be done by reflecting on the consequences of the no boundary condition for the state of one temporal extremity of the universe, considered in isolation. If that worked for what we call the initial state it would also work for what we call the final state—unless of course the argument had unwittingly assumed an objective distinction between initial state and final state, and hence applied some constraint to the former that it didn't apply to the latter. Hawking needs a more general argument, to the effect that disorder-disorder universes are impossible (or at least overwhelmingly improbable). He needs to show that almost all possible universes have at least one ordered temporal extremity—or equivalently, at most one disordered extremity.

As Hawking himself points out, it would then be legitimate to invoke a weak anthropic argument to explain why we regard the ordered extremity thus guaranteed as an *initial* extremity.[19] In virtue of its consequences for temporal asymmetry elsewhere in the universe, and the way in which intelligent creatures like ourselves rely on the thermodynamic asymmetry, it seems plausible that conscious observers are bound to regard this ordered extremity to the universe as lying in their past. This is very much like Boltzmann's use of anthropic reasoning, which we encountered in chapter 2.

The first possibility is thus that Hawking has such an argument, but hasn't told us what it is (probably because he does not see why it is so important).[20] As I see it, the other possibilities are that Hawking has made one of two mistakes (neither of them the mistake he claims to have made). Either his no boundary condition does exclude disorder at both temporal extremities of the universe, in which case his mistake was to change his mind about contraction leading to decreasing entropy; or the proposal doesn't exclude disorder at either temporal extremity of the universe, in which case his mistake is to think that the no boundary condition accounts for the low-entropy big bang.

I have done my best to examine Hawking's published papers in order to discover which of these three possibilities best fits the case. A helpful recent paper is Hawking's contribution to a major conference on the arrow of time which was held in Spain in 1991.[21] In this paper Hawking describes the process by which he and various colleagues applied the no boundary condition to the question of temporal asymmetry. He recounts how he and Halliwell "calculated the spectrum of perturbations predicted by the no boundary condition."[22] The conclusion was that "one gets an arrow of time. The universe is nearly homogeneous and isotropic when it is small. But it is more irregular, when it is large. In other words, disorder increases, as the universe expands."[23] I want to note in particular that at this stage Hawking doesn't refer to the stage of the universe in temporal terms—*start* and *finish,* for example—but only in terms of size. Indeed, as I noted, he points out correctly that the temporal perspective comes from us, and depends in practice on the thermodynamic arrow.

Hawking then tells us how he

> made what I now realize was a great mistake. I thought that the no boundary condition, would imply that the perturbations would be small whenever the radius of the universe was small. That is, the perturbations would be small, not only in the early stages of the expansion, but also in the late stages of a universe that collapsed again. ... This would mean that disorder would increase during the expansion, but decrease again during the contraction.

Hawking goes on to say how he was persuaded that this was a mistake, as a result of the objections raised by Page and Laflamme. (At the time, Laflamme was one of Hawking's graduate students.) He says he came to accept that

> [w]hen the radius of the universe is small, there are two kinds of solution. One would be an almost Euclidean complex solution, that started like the north pole of a sphere, and expanded monotonically up to a given radius. This would correspond to the start of the expansion. But the end of the contraction, would correspond to a solution that started in a similar way, but then had a long almost Lorentzian period of expansion, followed by a contraction to the given radius. ... This would mean that the perturbations would be small at one end of time, but could be large and non-linear at the other end. So disorder and irregularity would increase during the expansion, and would continue to increase during the contraction.[24]

Hawking then describes how he, Laflamme, and another graduate student, Glenn Lyons, have "studied how the arrow of time manifests itself in the various perturbation modes." He says that there are two relevant kinds of perturbation mode, those that oscillate and those that do not. The former will be essentially time symmetric, about the time of maximum expansion. In other words, the amplitude of perturbation, will be the same at a given radius during the expansion, as at the same radius during the contracting phase.[25]

The latter, by contrast,

will grow in amplitude in general. ... They will be small, when they come within the horizon during the expansion. But they will grow during the expansion, and continue to grow during the contraction. Eventually, they will become non linear. At this stage, the trajectories will spread out over a large region of phase space.[26]

It is the latter perturbation modes which, in virtue of the fact they are so much more common, lead to the conclusion that disorder increases as the universe recontracts.

Let's think about the last quotation. If it isn't an objective matter which end of the universe represents expansion and which contraction, and there is no constraint which operates simply in virtue of the radius of the universe, why should the perturbations ever be small? Why can't they be large at both ends, compatibly with the no boundary condition?

I haven't been able to find an answer to this crucial question in Hawking's papers. However, a striking feature of the relevant papers is that Hawking talks of showing that the relevant modes *start off* in a particular condition. Let me give some examples (with my italics, throughout). In an important paper written with Halliwell in 1985, for example, the authors say in the abstract that they "show ... that the inhomogeneous or anisotropic modes *start off* in their ground state."[27] Later in the paper they say,

We show that the gravitational-wave and density-perturbation modes obey decoupled time-dependent Schrödinger equations with respect to the time parameter of the classical solution. The boundary conditions imply that these modes *start off* in the ground state.

and, "We use the path-integral expression for the wave function ... to show that the perturbation wave functions *start out* in their ground states."[28]

Finally, in Hawking's own 1985 paper "Arrow of Time in Cosmology" (the paper in which he concedes his "mistake" in a note added in proof), he says this:

Thus at *early* times in the exponential expansion, i.e., when the Universe is small, the physical perturbation modes of the universe have their minimum excitation. The Universe is in a state that is as ordered and homogeneous as it can be consistent with the uncertainty principle.[29]

How are we to interpret these references to how the universe *starts off,* or *starts out,* or to the *early* universe? Do they embody an assumption that one temporal extremity of the universe is objectively its start? Presumably Hawking would want to deny that they do so, for otherwise he has simply helped himself to a temporal asymmetry at this crucial stage of the argument. (As I noted earlier, Hawking is in other places quite clear that our usual tendency to regard one end of the universe as the start is anthropocentric in origin, though related to the thermodynamic arrow. Like Boltzmann, Hawking takes the view that because they depend on the entropy gradient, sentient creatures are bound to regard the low-entropy direction as the past.)[30] But without the assumption that one temporal extremity of the universe is "really" the beginning, what is the objective content of Hawking's conclusion? Surely it can only be that the specified results obtain when the universe is *small*—as Hawking's own gloss has it, in the last passage quoted—in which case the argument must work at both ends of the universe, or at neither.

In other words, a crucial step in Hawking's argument seems to be to consider what the no boundary condition implies about one extremity of an arbitrary universe—about the extremity he wants to think of as the "start." But what can this possibly mean? If we have not yet established that an arbitrary universe has a single ordered extremity, it cannot be taken to mean "the extremity that intelligent creatures will think of as the start," because we haven't yet shown that an arbitrary universe has such an extremity (and only one). So there seems to be no coherent way to formulate a vital step in Hawking's argument, as he construes it. The only way to make the argument coherent is to take it to apply to any temporal extremity, but in this case the consequences of the no boundary condition will be symmetric: if one end of the universe has to be ordered, so must the other be.

It is important to appreciate that a symmetric version of Hawking's argument is not automatically ruled out by statistical considerations about the "likely" fate of a recollapsing universe. In changing his mind about the symmetric view, Hawking appears to be have been moved by what is essentially a statistical consideration: the fact that (as Page convinced him) most possible histories for the universe lead to a disordered collapse. However, the lesson of the gravitational symmetry argument was that in the absence of any prior justification for a temporal double standard, statistical arguments defer to boundary conditions. (The big bang is smooth *despite* the fact that it is as much the result of gravitational collapse as a big crunch would be, and this trumps any general appeal to the clumpiness of the end states of gravitational collapse.) Accordingly, it would have been open to Hawking to argue that Page's statistical considerations were simply overridden by the no boundary condition, treated as a symmetric constraint on the temporal extremities of the universe. Given that he does not argue this way, however, he needs to explain why analogous statistical arguments do not apply toward (what we call) the big bang.

It seems then that Hawking did make a mistake about temporal asymmetry, but not the one he thought he made. The real mistake lies in the ambiguity concerning the scope of the no boundary condition itself. In effect, the ambiguity traps Hawking between two errors: that of assuming that the universe has an objective start, and that of failing to see that in virtue of the fact that application of the no boundary condition depends only on *size,* it must apply equally at both temporal extremities of the universe.

In particular, then, Hawking seems to be in no position to exploit the loophole we described earlier in the section: he hasn't shown that asymmetric universes are the natural product of a symmetric theory, like left-handed and right-handed corkscrews from the same unbiased factory. On the contrary, in effect, he seems to have simply assumed the required asymmetry, by taking the no boundary condition to apply to only one end of an arbitrary universe. As I noted earlier,[31] this amounts to putting the asymmetry in "by hand." Ironically, Hawking himself applies the same criticism to Roger Penrose's proposal for accounting for the smoothness of the early universe. As we shall see, Penrose's proposal is explicitly asymmetric, and Hawking says that this means that "[i]n effect, one is putting in the thermodynamic arrow by hand."[32] In my view, the difference is mainly that Penrose says that this is what he is doing, while Hawking mistakenly claims to achieve something more.

This criticism should be put in perspective, however. Like Boltzmann a century before him, Hawking is better than many of his contemporaries at thinking in atemporal ways. (An illustration of this is Hawking's willingness to endorse the symmetric Gold view, at least initially. With less appreciation of the atemporal viewpoint, most cosmologists dismiss this view on what turn out to be quite fallacious grounds, as we shall see in a moment.) Again like Boltzmann, however, Hawking fails to apply the atemporal view in a consistent and thorough way. Had Boltzmann done so, his discussion of the asymmetry of the *H*-theorem in the 1890s would have taken a different path; had Hawking done so, he would surely have responded differently to the initial criticisms of his no boundary condition proposal.

THE BASIC DILEMMA AND SOME WAYS TO AVOID IT

The examples in the two previous sections suggest that even some of the most capable of modern cosmologists have difficulty in grasping what I called the basic dilemma of cosmology and time asymmetry. Yet this dilemma is really just a new manifestation of the problem that the physics of time asymmetry has been faced with since Boltzmann's time: How is it possible to derive the asymmetric world we find around us from the symmetric laws we find in physics? Transposed to a cosmological key, where the puzzle is to explain the low entropy of the big bang, this old challenge confronts us with a uncomfortable choice: it seems that we have to accept either that entropy must decrease toward a big crunch, as well as a big bang, or that the low-entropy big bang is simply not explicable by a time-symmetric physics.

Perhaps the only contemporary writer who fully appreciates the force of this dilemma is Roger Penrose, whose numerical estimate of the "specialness" of the big bang I mentioned earlier in the chapter. Penrose's own preference is for the second horn of the dilemma. Accordingly, he concludes that the underlying physics of the universe must be asymmetric in time—that there must be an additional *asymmetric* law of nature, to the effect that the initial extremities of the universe obey what amounts to a smoothness constraint.[33] (In technical terms, his hypothesis is that the so-called Weyl curvature of spacetime approaches zero in this initial region.) In effect, Penrose's argument is that it is reasonable to believe that such a constraint exists, because otherwise the universe as we find it would be unbelievably improbable.

Readers who have got the hang of the atemporal perspective might object at this point that the use of the term *initial* in the formulation of this hypothesis seems to violate the requirement that the initial/final distinction not be regarded as of any objective significance. Isn't Penrose already presupposing an objective distinction, in order to be able to formulate his hypothesis? In this case, however, the difficulty turns out to be superficial. Penrose's claim need only be that it is a physical law that there is one temporal direction in which the Weyl curvature always approaches zero toward the universe's extremities. The fact that conscious observers inevitably regard that direction as the past will then follow from the sort of weak anthropic argument already mentioned.

Notice, however, that even an advocate of the first horn of the dilemma— the time-symmetric Gold view—might be convinced by Penrose's argument that the observed condition of the universe can only be accounted for by a new physical law. The required law might be that the Weyl curvature approaches zero toward the extremities of the universe in both directions. Obviously this alternative would do just as well as Penrose's proposal at explaining the smoothness of the big bang, and it has the advantage of not introducing asymmetry into physics. If Penrose's option is to be preferred, there had better be some good reasons for rejecting the Gold view. As we shall see in the next section, however, most of the arguments that cosmologists give for rejecting the Gold view are very weak. This means that there is little or no reason to prefer Penrose's asymmetric hypothesis to the symmetric version of the same thing.[34]

Before we turn to the Gold view, which is the first horn of the basic dilemma, I want to mention some ways in which it might be possible to avoid the dilemma altogether.

What if there is no big crunch?

First of all, it might be thought that the basic dilemma only arises if the universe eventually recollapses. However, it is currently an open question in cosmology whether the universe contains enough matter to slow and reverse its own expansion. If the universe goes on expanding for ever, then it has no extremity toward (what we call) the future. In this case, an argument showing that all extremities are smooth and ordered would not imply Gold's view. Entropy would be free to go on increasing for ever.

This point is an interesting one, but it should not be overrated. For one thing, if we are interested in whether the Gold universe is a coherent possibility, the issue as to whether the actual universe recollapses is rather peripheral. The main issue is whether the Gold view makes sense in a recollapsing universe, not whether our universe happens to be a recollapsing universe. Of course, if we could show that a recollapsing universe is impossible, given the laws of physics as we know them, the situation would be rather different: we would have shown that the original puzzle concerns a case that physics allows us to ignore.

In practice, there does seem to be some prospect that recollapsing universes will turn out to be physically impossible. One of the features of the inflationary models we mentioned earlier in the chapter is that they predict that the universe will have a density extremely close to the critical value which must be exceeded if the universe is to recollapse. So the inflationary paradigm in cosmology might indeed have the consequence that the universe can never recollapse.

However, even if the universe as a whole never recollapses, it seems that parts of it do, as massive objects of various kinds form black holes. (It is now accepted that this is the normal fate of some massive stars, for example.) As many cosmologists have pointed out, this local process of gravitational collapse is like a miniature version of what the big crunch would be, if there were such a thing. More to the point, these local gravitational collapses seem to be smaller versions of the global gravitational collapse we see if we view the big bang in reverse. So the basic dilemma rears its horns once more: if a symmetric physics implies that the big bang must be smooth, then surely it will also imply that the collapse which gives rise to a black hole will result in a smooth final distribution of the collapsing material—an outcome as much in conflict with the natural "clumping" tendency of gravitating matter as a smooth big crunch would be. For this reason, writers such as Hawking and Penrose have concluded that the issue as to whether the universe as a whole recollapses is largely irrelevant to the problems of explaining the low-entropy big bang.[35]

Even if we could ignore the problem of black holes, however, the asymmetry of a universe which never recollapses would be of no immediate help in explaining the smoothness of what we regard as the early stages of the universe. This is because from the objective atemporal standpoint, there would be no reason to prefer the usual description of such a universe—as one which is expanding indefinitely from a big bang—to the reverse description, which sees it as a universe which has always been contracting toward a big crunch. The basic difficulty is that from the atemporal standpoint there is no objective reason to characterize what we think of as the big bang as the start of an expansion, rather than the endpoint of a gravitational collapse; and so it is mysterious why it lacks the expected clumpiness of a gravitational collapse. This point is unaffected by whether there is a second extremity in the opposite temporal direction.

In a sense, then, the Gold universe is just a convenient way of thinking about a problem which arises independently. The intrinsic symmetry of a recollapsing universe ensures that without leaving the comfort of our ordinary temporal perspective, we are confronted with the question which the atemporal perspective requires us to ask about the big bang itself: How could such an event possibly have the properties it must have, in order to account for what we observe around us. Gold's contribution is to identify a possibility we must take seriously, if we are to ask this question without the illusory comfort of a temporal double standard.

The corkscrew model: asymmetric models for symmetric theories

Perhaps the best prospect for avoiding the basic dilemma is the loophole I mentioned earlier, in my discussion of Hawking's proposal. We saw that it is possible for a symmetric theory to have only (or mostly) asymmetric models. I suggested the analogy of a factory which produces equal numbers of right-handed and left-handed corkscrews: each corkscrew is spatially asymmetric, but the production as a whole shows no such bias. This seems an attractive solution to the basic dilemma, at least in comparison to the alternatives, but two notes of caution seem in order.

First, as I pointed out earlier, a proposal of this kind should not simply put in the required asymmetry "by hand." It must imply that all solutions are asymmetric, without simply assuming that this is the case. It is difficult to lay down precise guidelines for avoiding this mistake—after all, if the required asymmetry were not already implicit in the theoretical premises in some sense, it could not be derived from them—but presumably the asymmetry should flow from principles which are not *manifestly* asymmetric, and which have independent theoretical justification.

Second, we should not be misled into *expecting* a solution of this kind by the sort of statistical reasoning employed in first step of the gravitational symmetry argument. In particular, we should not think that the intuition that the most likely fate for the universe is a clumpy gravitational collapse makes a solution of this kind prima facie more plausible than a globally symmetric model of Gold's kind. The point of the gravitational symmetry argument was that these statistical grounds are temporally symmetric: if they excluded Gold's suggestion, then they would also exclude models with a low-entropy big bang. In effect, the hypothesis that the big bang is explicable is the hypothesis that something—perhaps a boundary condition of some kind, perhaps an alternative statistical argument, conducted in terms of possible models of a theory—defeats these statistical considerations in cosmology. As a result, we are left with no reason to *expect* an asymmetric solution in preference to Gold's symmetric proposal. On the contrary, the right way to argue seems to be something like this: the smoothness of the big bang shows that statistical arguments based on the character of gravitational collapse are not always reliable—on the contrary, they are unreliable in the one case (out of a possible two!) in which we can actually subject them to observational test. Having discovered this, should we continue to regard them as reliable in the remaining case (i.e., when oriented toward the big crunch)? Obviously not, at least in the absence of any independent reason for applying such a double standard.

Things would be different if we were prepared to allow that the low-entropy big bang is *not* explicable—that it is just a statistical "fluke." In this case we might well argue that we have very good grounds to expect the universe to be "fluky" only at one end. However, at this point we would have abandoned the strategy of trying to show that almost all possible universes are asymmetric, the goal of which was precisely to *explain* the low-entropy big bang. Instead we might be pursuing a different strategy altogether ...

The anthropic strategy

Perhaps the reason that the universe looks so unusual to us is simply that we can only exist in very unusual bits of it. We depend on the entropy gradient, and could not have evolved or survive in a region in thermodynamic equilibrium. Perhaps this explains why we find ourselves in a region of the universe exhibiting such a gradient.

This is the anthropic approach, which we already encountered in the form of Boltzmann's idea that we live in the kind of incredibly rare statistical fluctuation that the statistical approach to thermodynamics allows. As in Boltzmann's time, the idea is an interesting one, but has to face up to some severe difficulties. The first is that it depends on there being a genuine multiplicity of actual "bits" of a much larger universe, of which our bit is simply some small corner. It is no use relying on other merely *possible* worlds, since that would leave us without an explanation for why ours turned out to be the actual world.[36] (If it hadn't turn out this way, we wouldn't have been around to think about it, but this doesn't explain why it did turn out this way.) So the anthropic solution is exceedingly costly in ontological terms—that is, in terms of what it requires that there be in the world. In effect, it requires that there be vastly more "out there" than we are ordinarily aware of—even as long-range astronomers!

Penrose's numerical estimate of the unlikeliness of the smooth big bang gives us some indication of the scale of this problem. According to Penrose's estimate, only 1 in $10^{10^{123}}$ of universes will have the right sort of big bang. This means that the anthropic strategy has to take seriously the possibility that our universe comprises something like this incredibly tiny fraction of the whole of reality. All the same, this would not be a disadvantage if the cost was one we were committed to bearing anyway. It turns out that according to some versions of the inflationary theory in contemporary cosmology, universes in the normal sense are just bubbles in some vast foam of universes. So there might be independent reason to believe that reality is vastly more inclusive than it seems. In this case, the anthropic view does not necessarily make things any worse.

The second main difficulty for the anthropic strategy is that as Penrose himself emphasizes,[37] there may well be much less costly ways to generate a sufficient entropy gradient to support life. Penrose argues that the observed universe is vastly more unlikely than intelligent life requires. Again, this is close to an objection to Boltzmann's view. I noted that Boltzmann's suggestion implies that at any given stage, we should not expect to find more order than we have previously observed. As we look farther into space, in particular, we should not expect to find more and more galaxies and suns like our own. A fluctuation which produces one galaxy is more than adequate for human life, and vastly cheaper in entropy terms than a fluctuation which produces two galaxies, let alone two million, or two billion. The same seems to apply to the contemporary argument: life as we know it doesn't seem to require an early universe which is smooth everywhere, but only one which is smooth in a sufficiently large area to allow a galaxy or two to form (and to remain relatively undisturbed while intelligent life evolves). This would be much cheaper in entropy terms than global smoothness.

However, the inflationary model might leave a loophole here, too. If the inflationary theory could show that a universe of the size of ours is an all or nothing matter, then the anthropic argument would be back on track. The quantum preconditions for inflation might be extremely rare, but this would not matter, so long as (1) there is enough time in some background grand universe for them to be likely to occur eventually, and (2) it is guaranteed that when they do occur a universe of our sort arises, complete with its smooth boundary.

Hence it seems to me that the anthropic strategy does provide a possible escape from the basic dilemma. It depends heavily on the right sort of assistance from cosmological theory, but if this were forthcoming, the anthropic approach could turn out to explain why we find ourselves in a universe with a low-entropy history—without implying that there must be a low-entropy future, if the universe recollapses. If so, however, then there is hugely more to reality than we currently imagine, and even the vast concerns of contemporary astronomy will pale into insignificance in comparison.

This is a possible solution, then, but in human terms a far less welcome one than would be the discovery that our universe could be explained as it stands, as a natural product of physical laws. The basic dilemma stems from the fact that the available laws seem to be time-symmetric, and hence likely to constrain both extremities of the universe in the same way. Let us now turn to the reasons why this has usually been regarded as such an absurd idea, since Gold first suggested it nearly forty years ago.

WHAT'S WRONG WITH A GOLD UNIVERSE?

A Gold universe has low entropy at both ends. In a sense, then, it leaves us with two special boundary conditions to explain, rather than just one. Of course, it makes up for doubling the problem by vastly improving the prospects that a symmetric physics will be up to the explanatory task. As we saw earlier, Gold himself was attracted to the idea that it is the expansion itself which ensures that entropy is high when the universe is large, and low when the universe is small. His initial thought was that the expansion itself increases the maximum possible entropy, by creating new possible configurations for matter. (The local entropy-increasing processes we observe would just represent some of the many ways in which the contents of the universe come to occupy these newly available niches.)

This suggestion soon turned out to be untenable, however. Cosmologists saw that contraction itself would not be sufficient to make the contracting half of a recollapsing universe a kind of mirror image of the expanding half as we know it—to make radiation converge on stars, for example.[38] Later, this point was underscored by the realization that smoothness is a crucial aspect of the low-entropy big bang. As we saw, smoothness is anything but a natural product of gravitational contraction. So the clue to the low-entropy extremities of a Gold universe does not seem to lie in the process of expansion and contraction itself. The extremities need to be special, in a way which expansion and contraction alone do not explain.

One option would be to accept the low entropy of the temporal extremities of the universe as an additional law of nature. As I mentioned earlier, this would be in the spirit of Penrose's proposal, but it would still be a time-symmetric law. True, we might find such a law somewhat ad hoc. But this might seem a price worth paying, if the alternative is that we have no explanation for such a striking physical anomaly. And the proposal would seem a lot less ad hoc if it could be grounded on attractive theoretical considerations of some kind—perhaps Hawking's no boundary condition, in a symmetric version, for example.

But what of the consequences of a Gold universe? If the second law of thermodynamics changes direction when the universe recontracts, the universe would enter an age of apparent miracles. Radiation would converge on stars, apples would compose themselves in decompost heaps and leap into trees, and humanoids would arise from their own ashes, grow younger, and become unborn. However, by now it should be obvious that such apparently miraculous behavior does not constitute an objection to this symmetric model of the universe, on pain of the old double standard. This point is made rather nicely by Paul Davies. After describing some "miraculous" behavior of this kind, Davies continues:

> It is curious that this seems so laughable, because it is simply a description of our present world given in reversed-time language. Its occurrence is *no more remarkable* than what we at present experience—indeed it *is* what we actually experience—the difference in description being purely semantic and not physical.[39]

Davies goes on to point out that the difficulty really lies in managing the transition: "What *is* remarkable, however, is the fact that our 'forward' time world *changes into* [a] backward time world (or vice versa, as the situation is perfectly symmetric)."

What exactly are the problems about this transition? In the informal work from which I have just quoted, Davies suggests that the main problem is that it requires that the universe have very special initial conditions.

> Although the vast majority of microscopic motions in the big bang give rise to purely entropy-increasing worlds, a very, very special set of motions could

indeed result in an initial entropy increase, followed by a subsequent decrease. For this to come about the microscopic constituents of the universe would not be started off moving randomly after all, but each little particle, each electromagnetic wave, set off along a carefully chosen path to lead to this very special future evolution. ... Such a changeover requires ... an extraordinary degree of cooperation between countless numbers of atoms.[40]

Davies here alludes to his earlier conclusion that "a time-asymmetric universe does not demand any very special initial conditions. It seems to imply a creation which is of a very general and random character at the microscopic level."[41] However, we have seen that to maintain this view of the early universe while invoking the usual statistical arguments with respect to the late universe is to operate with a double standard: double standards aside, the gravitational symmetry argument shows that if a late universe is naturally clumpy, so too is an early universe. In the present context the relevant point is that (as Davies himself notes, in effect)[42] the conventional time-asymmetric view itself requires that the final conditions of the universe be microscopically arranged so that when viewed in the reverse of the ordinary sense, the countless atoms cooperate over billions of years to achieve the remarkable low-entropy state of the big bang. Again, therefore, a double standard is involved in taking it to be an argument against Gold's view that it requires this cooperation in the initial conditions. As before, the relevant statistical argument is an instrument with two possible uses. We know that it yields the wrong answer in one of these uses, in that it would exclude an early universe of the kind we actually observe. Should we take it to be reliable in its other use, which differs only in temporal orientation from the case in which the argument so glaringly fails? Symmetry and simple caution both suggest that we should not.

A different sort of objection to the Gold view rests on thought experiments concerning objects which survive from one half of the Gold universe to the other. These thought experiments are made a lot more realistic by the fact that the Gold view seems committed to saying that entropy decreases not just toward a big crunch, but also toward its more localized analogs, namely the localized gravitational collapses which are now thought to form black holes. (Recall that in the previous section, this was one of the reasons for saying that we don't escape the basic dilemma if the universe as a whole never recollapses.) Accordingly, in the following passage Roger Penrose describes what he takes to be an unacceptable consequence of the Gold model:

Let us envisage an astronaut in such a universe who falls into a black hole. For definiteness, suppose that it is a hole of 10^{10} [solar masses] so that our astronaut will have something like a day inside [the event horizon], for most of which time he will encounter no appreciable tidal forces and during which he could conduct experiments in a leisurely way. ... Suppose that experiments are performed by the astronaut for a period while he is inside the hole. The behaviour of his apparatus (indeed, of the metabolic processes within his own body) is entirely determined by conditions at the black hole's singularity ...— as, equally, it is entirely determined by the conditions at the big bang. The situation inside the black hole differs in no essential respect from that at the late stages of a recollapsing universe. If one's viewpoint is to link the local direction of time's arrow directly to the expansion of the universe, then one must surely be driven to expect that our astronaut's experiments will behave in an entropy-*decreasing* way (with respect to "normal" time). Indeed, one should presumably be driven to expect that the astronaut would believe himself to be coming out of the hole rather than falling in (assuming his metabolic processes could operate consistently through such a drastic reversal of the normal progression of entropy).[43]

However, I think this argument rests on a mistake about the nature of the temporal reversal in these time-symmetric models. Consider Penrose's astronaut. He is presumably a product of a billion years of biological evolution, to say nothing of the ten billion years of cosmological evolution which created the conditions for biology to begin on our planet. So he is the sort of physical structure that could only exist at this kind of temporal distance from a suitable big bang. What counts as suitable? The relevant point is that low entropy doesn't seem to be enough; for one thing, the "bang" will need to be massive enough to produce the cosmological structure on which life depends.

This means that Penrose's astronaut is not going to encounter any time-reversed humanoids inside the black hole, or any unevolving life, or even any unforming stars and galaxies. More important, it means that he himself has no need of an inverse evolutionary history inside the hole, in addition to the history he already has outside. He need not be a "natural" product of the hole's singularity. Relative to its reversed time sense, he is simply a "miracle"—an incredibly unlikely chance event. The same goes for his apparatus—in general, for all the "foreign" structure he imports into the hole.

Notice that there are two possible models of the connections that might hold between the products of two low-entropy boundary conditions. The first is a "meeting" model, in which any piece of structure or order is a "natural" product of singularities or extremities in both temporal directions. The second is a "mixing" model, in which structure or order is normally a

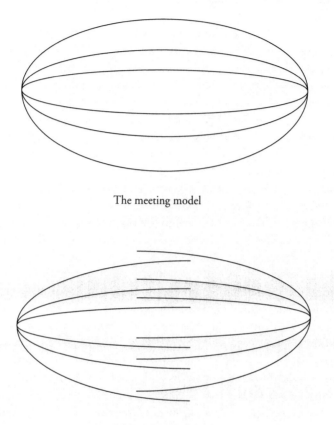

The meeting model

The mixing model

Figure 4.1. Two models for a Gold universe.

product of an extremity in one direction or the other, but not usually both. (These two different models are represented in Figure 4.1.)

Penrose's argument seems to take for granted that the meeting model is the right one to apply in thinking about Gold's time-symmetric view. Elsewhere, Stephen Hawking also seems to assume this model; he suggests that the astronaut entering the event horizon of a black hole wouldn't notice the time reversal because his psychological time sense would reverse.[44] However, the meeting model seems to place a quite unnecessary constraint on the Gold view. The right guiding principle seems to be that any piece of low-entropy "structure" needs to explained *either* as a product of a past singularity *or* as a product of the future singularity; but that no piece needs both sorts of explanation.

The proportions of each kind can be expected to vary from place to place. In our region of the universe, relatively close to the big bang, and far from black holes, virtually all the structure results from the big bang. This might continue to be the case if in the future we fall into the sort of black hole which doesn't have the time or the mass to produce much structure of its own. In this case the experience might be very much less odd than Penrose's thought experiment would have us believe. The reverse structure produced by the black hole might be insignificant for most of the time we survived within its event horizon.

What if we approach a black hole which is big enough to produce interesting structure—the big crunch itself, for example? Does Penrose's argument not still apply in this case? It seems to me that the case is still far from conclusive, so long as we bear in mind that *our* structure doesn't need a duplicate explanation from the opposite point of view. It is true that in this case we will expect eventually to be affected by the reverse structure we encounter. For example, suppose that our spacecraft approaches what from the reverse point of view is a normal star. From the reverse point of view we are an object leaving the vicinity of the star. We appear to be heated by radiation from the star, but to be gradually cooling as we move farther away from the star, thus receiving less energy from it, and radiating energy into empty space.

What would this course of events look like from our own point of view? Apparently we would begin to heat up as photons "inexplicably" converged on us from empty space. This inflow of radiation would increase with time. Perhaps even more puzzlingly, however, we would notice that our craft was reradiating toward one particular direction in space—toward what from our point of view is a giant radiation sink. Whether we could detect this radiation directly is a nice question—more on this below—but we might expect it to be detectable indirectly. For example, we might expect that the inside of the wall of the spaceship facing the reverse star would feel cold to the touch, reflecting what in our time sense would be a flow of heat energy toward the star.

These phenomena would certainly be bizarre by our ordinary standards, but it is not clear that they constitute an objection to the possibility of entropy reversal. After all, within the framework of the Gold model itself they are not in the least unexpected or inexplicable. To generate a substantial objection to the model, it needs to be shown that it leads to incoherencies of some kind, and not merely to the unexpected. Whether this can be shown seems to be an open question, which I want to discuss further in the next section.

Penrose himself no longer puts much weight on the astronaut argument, saying that he now thinks that a much stronger case can be made against the suggestion that entropy decreases toward singularities. He argues that in virtue of its commitment to temporal symmetry this view must either disallow black holes in the future, or allow for a proliferation of white holes in the past. He says that the first of these options "requires physically unacceptable teleology," while the second would conflict with the observed smoothness of the early universe.[45] However, the objection to the first option is primarily statistical: "it would have to be a seemingly remarkably improbable set of coincidences that would forbid black holes forming. The hypothesis of black holes being not allowed in the future provides 'unreasonable' constraints on what matter is allowed to do in the past."[46] And I think this means that Penrose is again invoking a double standard, in accepting the "naturalness" argument with respect to the future but not the past. Once again: the lesson of the smooth past seems to be that in that case something overrides the natural behavior of a gravitational collapse; once this possibility is admitted, however, we have no non–question-begging grounds to exclude (or even to *doubt!*) the hypothesis that the same overriding factor might operate in the future.[47]

As it stands, then, the arguments against the Gold view do not seem very convincing. Some of them involve the kind of temporal double standard which has cropped up so often in discussions we have been looking at in this and the two previous chapters. The most interesting line of argument concerned the possible interactions between the two halves of a Gold universe. In the next section I want to explore these issues a little further. As I want to show, they raise the interesting possibility that in principle, astronomers in either half of a Gold universe might be able to detect some of the features of the other half—even though those features lie in the distant future, from the astronomers' point of view.

A TELESCOPE TO LOOK INTO THE FUTURE?

Let us suppose that our universe is actually a Gold universe: it eventually recollapses, and the contracting phase (as we call it) is much like the expanding phase, with the familiar temporal asymmetries reversed. The contracting phase will contain what in our time sense look like reverse galaxies and reverse stars. These are exactly like our stars and galaxies, but with the opposite time sense. In our time sense, they are sinks for radiation, rather than sources. (Coherent radiation converges on them, rather than diverging from them.)

What happens if we point one of our telescopes in the direction of one of these reverse galaxies—in other words, if we point the telescope in what happens to be the right direction to receive some of the light which, from the standpoint of astronomers in the reverse galaxy, left their galaxy billions of years previously? What effect, if any, would this have on our telescope and its surroundings?

A useful first step is to consider matters from the reverse point of view. An astronomer in the reverse galaxy will say that light emitted from her galaxy is being collected and absorbed by the distant telescope on Earth. (She won't know that this is happening, of course, because from her point of view it is taking place far in the future. There is nothing to stop her speculating about the possibility, however.) Accordingly, she will expect certain effects to take place at the back of the telescope, due to the light's absorption. A black plate placed at the back of the telescope will be heated slightly by the incoming radiation, for example.

What does this look like from our point of view? Our temporal sense is the reverse of that of the distant astronomer, so that what she regards as absorption of radiation seems to us to be emission, and vice versa. Similarly, apparent directions of heat flow are also reversed. Thus as we point our telescope toward the distant reverse galaxy, the effect should be a sudden increase in the flow of radiation from the telescope into space; and, indirectly, an apparent *cooling* of the black plate at the rear of our telescope. Why cooling and not heating? Because heat is flowing into the plate from its surroundings, in out time sense, and then away into space as light radiation. The plate is actually hotter than its immediate environment, but it behaves in the way we normally expect of an object which is cooler than its surroundings: in other words, it takes in heat from its environment.[48]

As in normal astronomy, the size of these effects will depend on the distance and intensity of the reverse source in question. In practice, the interval between our era and the corresponding era in the contracting phase of a Gold universe might be so vast that any effects of this kind would be insignificant. These practical difficulties should not prevent us from exploring these ideas in principle, however. For one thing, we might manage to turn up some sort of logical difficulty, which would rule out the Gold universe once and for all. (Many thought experiments in physics are impossible to perform, but hardly less important on that account.)

The size of the effects aside, there seem to be theoretical difficulties in detecting them by what might seem the obvious methods. For example, it will be no use placing a photographic plate over the aperture of the telescope, hoping to record the emission of the radiation on its way to the reverse galaxy. If we consider things from the point of view of the distant reverse astronomer, it is clear that the plate would act as a shield, shading the telescope from the light from her galaxy. Thus from our point of view the light will be emitted from the back of the plate—from the side facing away from the telescope, toward the reverse galaxy.

Let us look at the expected behavior of the telescope in a little more detail. When we shine a light at an absorbing surface we expect its temperature to increase. If the incoming light intensity is constant the temperature will soon stabilize at a higher level, as the system reaches a new equilibrium. If the light is then turned off the temperature drops exponentially to its previous value. Hence if future reverse astronomers shine a laser beam in the direction of one of our telescopes, at the back of which is an absorbing plate, the temperature change they would expect to take place in the plate is as shown in Figure 4.2a. When the telescope is opened the temperature of the plate rises due to the effect of the incoming radiation, stabilizing at a new higher value. If the telescope is then shut, so that the plate no longer absorbs radiation, its temperature drops again to the initial value.

Figure 4.2b shows what this behavior looks like from our point of view. Setting aside the issue mentioned above of the *apparent* temperature of the plate relative to its surroundings, the only change is in the temporal ordering of the relevant events. One of the striking things about this behavior is that it appears to involve what physicists call advanced effects—effects which take place *before* the event which causes them. The temperature rises before we open the telescope, and falls before we close it. This suggests that we might be able to argue that the whole setup is incoherent, using the kind of argument often used to try to show that backward causation leads to paradoxical results. (We will look at these arguments in more detail in chapter 7.) Couldn't we adopt the following policy, for example: *Open the telescope only if the temperature of the black plate has not just risen significantly above that of its surroundings?* It might seem that this entirely feasible policy generates contradictory predictions, thus providing a reductio ad absurdum of the time-reversing view.

But are the results really contradictory? Grant for the moment that while this policy is in force it will not happen that the temperature of the plate rises on an occasion on which we might have opened the telescope, but didn't actually do so. This leaves the possibility that on all relevant trials the temperature does not rise, and the telescope is opened. Is this inconsistent with the presence of radiation from the future reverse source?

I don't think so. We should keep in mind that the temperature profile depicted in these diagrams relies on statistical reasoning: it is inferred from the measured direction of heat flow, and simply represents the most likely way for the temperature of the absorbing plate to behave. But one of the lessons of our discussion has been that statistics may be overridden by boundary conditions. Here, the temperature is constant before the telescope is opened

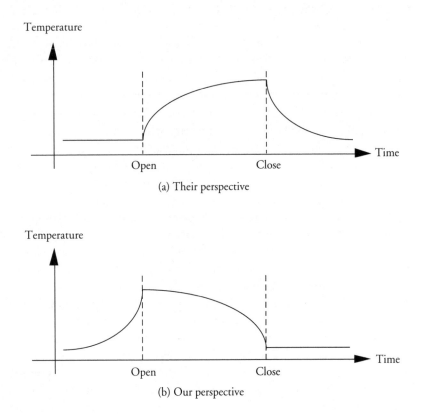

Figure 4.2. Two views of a telescope to look into the future.

because our policy has imposed this as a boundary condition. A second boundary condition is provided by the presence of the future reverse radiation source. Hence the system is statistically constrained in both temporal directions. We should not be surprised that it does not exhibit behavior predicted under the supposition that in one direction or other, it has its normal degrees of freedom. It is not clear whether this loophole will always be available, but my suspicion is that it will be. If nothing else, quantum indeterminism is likely to imply that it is impossible to sufficiently constrain the two boundary conditions to yield an outright contradiction.

A related objection to the Gold universe has recently been raised by the physicists Murray Gell-Mann and James Hartle. Gell-Mann and Hartle consider the present consequences of assuming that the universe produces stars and galaxies at both ends, in the way we have been discussing.

Consider the radiation emitted from a particular star in the present epoch. If the universe is transparent, it is likely to reach the final epoch without being absorbed or scattered. There it may either be absorbed in the stars or proceed past them toward the final singularity. If a significant fraction of the radiation proceeds past, then by time-symmetry we should expect a corresponding amount of radiation to have been emitted from the big bang. Observation of the brightness of the night sky could therefore constrain the possibility of a final boundary condition time-symmetrically related to the initial one.[49]

In other words, the argument is that the Gold universe implies that there should be more radiation observable in the night sky than we actually see. As well as the radiation produced by the stars of our own epoch, there should be radiation which in the reverse time sense is left over from the stars of the reverse epoch. As Paul Davies and Jason Twamley describe Gell-Mann and Hartle's conclusion, "by symmetry this intense starlight background should also be present at our epoch ..., a difficulty reminiscent of Olbers' paradox."[50]

But if there were such additional radiation of this kind in our region of the universe, could we actually detect it? Gell-Mann and Hartle overlook this issue. The problem is that the radiation concerned is "already" neatly arranged to converge on its future sources, not on our eyes or instruments. Imagine, for example, that a reverse galaxy in direction +x is emitting (in its time sense) toward a distant point in direction –x (see Figure 4.3). We stand at the origin, and look toward –x. Do we see the light which in our time sense is traveling from –x toward +x? No, because we are standing in the way! If we are standing at the origin (at the relevant time) then the light emitted from the reverse galaxy falls on us, and never reaches what we think of as the past sky. When we look toward –x, looking for the radiation converging on the reverse galaxy at +x, then the relevant part of the radiation doesn't come from the sky in the direction –x at all; it comes from the surface at the origin which faces +x—that is, from the back of our own head! As in the telescope case, then, we discover that the radiation associated with the opposite end of a Gold universe is not necessarily detectable by normal means.

Thus the whole issue of the consequences and consistency of the Gold view is a lot more complicated than it looks at first sight. One of the general lessons is that because our ordinary (asymmetric) ways of thinking are intimately tied up with the thermodynamic asymmetry, we cannot assume that they will be dependable in contexts in which this asymmetry is not universal. To give a simple example, suppose that an event B follows deterministically from an event A. In a Gold universe we may not be able to say that if A had not happened B would not have happened—not because there is some alternative earlier cause waiting in the wings if A fails to materialize (as happens

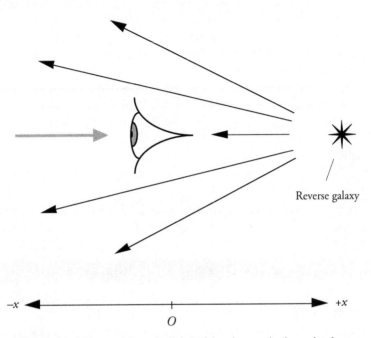

Figure 4.3. How not to see the light? If the observer looks to the sky in direction −*x*, hoping to see light which in her time sense would appear to be converging on the reverse galaxy in direction +*x*, she herself shades the sky in that direction, and so sees no light.

in cases of what philosophers call preemptive causation, for example), but simply because B is guaranteed by *later* events.

Figure 4.3 illustrates a consequence of this kind. We had a choice as to whether to interpose our head and hence our eye at the point *O*. If we had not done so, the light emitted (in the reverse time sense) by the reverse galaxy at +*x* would have reached −*x*, *in our past*. Our action thus influences the past. Because we interpose ourselves at *O*, some photons are not emitted from some surface at −*x*, whereas otherwise they would have been. Normally claims to affect the past give rise to causal loops, and hence inconsistencies. But again it is not obvious that this will happen in this case, for reasons similar to those in the telescope case.

These issues call for a lot more thought, but we can draw two rather tentative conclusions. First, the question whether Gold's view leads to some kind of inconsistency is still open (and won't be settled until we learn to think about the problem in the right way). Second, there is some prospect that the

contents of the contracting half of a Gold universe might be presently observable, at least in principle—despite the fact that they lie in what we think
of as the distant future.[51] The methods involved look bizarre by ordinary
standards, but in the end this is nothing more than the apparent oddity of
perfectly ordinary asymmetries having the reverse of their "usual" orientation. One of the main lessons of the last three chapters is that until we have
learned to disregard that sort of oddity, we will make no progress at all with
the problem of explaining temporal asymmetry.

CONCLUSION

At the beginning of the chapter I noted that one of the outstanding achievements of modern cosmology has been to offer some prospect of an answer
to the century-old mystery as to why entropy is so low in the past. In reveal-
ing the importance of the smooth early universe, contemporary cosmology
allows Boltzmann's great puzzle to be given a concrete form: Why is the
universe in this highly ordered condition, early in its history? Cosmology
thus inherits the project that Boltzmann began.

When it comes to the conduct of this project, however, we have found
that even some of the most able of contemporary cosmologists seem to have
trouble grasping the nature of the problem. They are prone to the same temporal double standards that have always afflicted debates on time asymmetry
in physics. As a result, most writers in this field have not seen the force of
the basic dilemma. That is, they have not appreciated how difficult it is to
explain why the universe is smooth at one end, without at the same time
showing that it must be smooth at the other (so that the familiar arrows of
time would reverse if the universe recontracts).

Despite the insights of modern cosmology, then, Boltzmann's project is far
from completion. Until the ground rules are understood by the people who
are now qualified to play, further progress seems unlikely. In this chapter I
have tried to point out some of the characteristic mistakes in the game as it is
currently played, in the hope of encouraging a more productive attack on the
remaining mysteries of temporal asymmetry. I want to finish by summarizing
the options and prospects for a satisfactory explanation of the smooth early
universe, as they seem in light of the discussion earlier in the chapter.

One attractive solution would be the possibility I described in terms of the
corkscrew model—in other words, a demonstration that although the laws
that govern the universe are temporally symmetric, the universes that they
allow are mostly asymmetric; mostly such that they possess a single temporal
extremity with the ordered characteristics of what we call the big bang.

However, it is important not to regard this option as an attractive one for the wrong reason. Cosmologists like it because it avoids what they see as the objections to the time-reversing Gold universe. But most of these objections rest on the usual statistical considerations—they rely on pointing out that the Gold view requires "unlikely" events of some kind. With double standards disallowed, however, these statistical arguments work equally well, or equally badly, in both directions. If they don't rule out a smooth big bang, then they don't rule out a smooth big crunch. So if anything it is the Gold view which should be regarded as the more plausible option, simply on symmetry grounds—at least in the absence of properly motivated objections to time-reversing cosmologies.

Failing either of these approaches, the main option seems to be an anthropic account, which claims that although there are lots of universes, most of which are unlike ours—they don't have a smooth big bang, for example—only a universe like ours can support intelligent life. True, there is also Penrose's view, which invokes a special asymmetric principle to explain the smooth big bang. But here the asymmetry seems rather ad hoc. If we are going to invoke a new principle, we might as well have a symmetric one, at least in the absence of decisive objections to the Gold view.

Thus we have four main alternatives: the corkscrew view, the Gold universe, the anthropic approach, and Penrose's asymmetric law proposal. To see how these views compare, let us think about how they regard the big bang. We saw earlier that the best way to appreciate the "specialness" of a smooth big bang is to think of it, with the usual temporal perspective reversed, as the end point of a gravitational collapse. The puzzle stems from the fact that such an endpoint would be expected to be clumpy, rather than smooth.

In effect, the anthropic view says the big bang can be thought of as a natural product of gravitational collapse—though as one of the naturally occurring "freak" outcomes, which are bound to occur every so often. How often? If we take Penrose's calculation as a guide, then the big bang is the kind of collapse we should expect to get once every $10^{10^{123}}$ attempts. As we saw, the anthropic approach thus requires that reality as a whole be vastly larger than what we know as the observable universe. Reality has to roll the dice often enough for an outcome as unlikely as this to occur with reasonable probability.

The other three options all say that the big bang cannot be thought of as a natural product of gravitational collapse. All of them say, in effect, that the kind of reasoning we use when we think about the normal result of a gravitational collapse is inapplicable to the case of the big bang. In one way or another, something overrides the considerations we invoke when we think about generic gravitational collapse.[52] Somehow, additional boundary conditions, or global constraints on the nature of the models a theory allows, manage to ensure that nature operates to much more restrictive rules. (What counts in favor the Gold view is that once we make this concession at one extremity, there seems to be no objection in principle to applying it more widely: no objection in principle, and yet the considerable advantages of symmetry.)

The anthropic view aside, then, cosmology seems to confirm one of the underlying lessons of the debate about the thermodynamic arrow. For all their apparent force, statistical arguments are surprisingly powerless. Double standards disallowed, the fact that these arguments don't work toward the past implies that we have no reason to expect that they will work toward the future. I want to finish by noting that this adds an interesting twist to the fundamental puzzle of temporal asymmetry, the question why entropy is low in the past.

We saw that Boltzmann's statistical treatment of entropy in thermodynamics makes it very puzzling why the entropy of our surroundings should be so low, compared to its theoretical maximum; and even more puzzling why entropy was even lower in the past. But what *exactly* is the problem here? Apparently, just that a low-entropy past is exceedingly unlikely *in statistical terms.* However, if one of the lessons of the debate is that statistical reasoning is unreliable, doesn't this undercut our reason for thinking that the low-entropy past is unusual, and hence in need of explanation?

This little twist shows just how difficult it is to be virtuous about the atemporal standpoint. The right conclusion to draw seems to be something like this: The smooth early universe does need to be explained, but not really on the grounds that it is statistically unlikely. It needs to be explained because it is such a distinctive and important aspect of the development of the universe as we know it—essential, apparently, to much else that we want to explain, such as the formation of galaxies, and the production of more familiar kinds of temporal asymmetry. This way of putting the problem avoids the questionable reliance on the statistical viewpoint, but leaves plenty of work for cosmologists!

5

⟵────⟶

Innocence and Symmetry
in Microphysics

THE three previous chapters looked at the three main ways in which the
arrow of time appears in modern physics. We considered the arrows of
thermodynamics, radiation, and cosmology, and I showed how the project
of trying to understand and explain these arrows has been riddled with some
very persistent mistakes. The most basic mistake is the double standard fallacy, which occurs when the two directions of time are unwittingly treated
in quite different ways. As an antidote to these mistakes, I recommended
the atemporal Archimedean standpoint, which is explicit in treating the past
and the future in an even-handed way. The lesson is that whenever physics
is confronted with the arrow of time, it should plan its response from this
detached viewpoint—"from nowhen," as I put it in chapter 1.

Important as these conclusions may be to the understanding of the physics
of time asymmetry, however, they may seem of little relevance elsewhere in
physics. The problem of the arrow of time is a rather specialized concern, on
the face of it, and so the detached atemporal viewpoint might seem of little
significance to mainstream issues. In the remainder of the book I want to
show that this impression is mistaken. The Archimedean standpoint turns
out to have profound implications elsewhere in physics.

In particular, the atemporal standpoint brings into focus a deep presupposition of contemporary physics, which is otherwise so natural as to be
almost invisible. Natural or not, this presupposition turns out to be highly
questionable, largely because it is temporally asymmetric. At the same time,
it turns out to play a crucial role in all the standard views of what quantum
mechanics—the most successful and yet most puzzling of modern physical
theories—is telling us about the world. Without the assumption in question,
quantum mechanics looks a lot less bizarre. By dropping the assumption,
then, we seem to gain on both the swings and the roundabouts: we get rid

of a principle whose temporal asymmetry is otherwise very puzzling, and we make a major dent in the huge mysteries of quantum mechanics itself.

The rest of the book aims to show that this rather astounding possibility is not some kind of metaphysical mirage. There really is a new pathway here, which, because the assumption concerned is so natural, so "intuitive," generations of thinkers have simply failed to see. The pathway offers a radically new view of the significance of quantum theory, a view which strips it of many of its most puzzling features. Whether it is the right way to understand quantum mechanics remains to be seen. However, I want to show that it is a lot more promising than anything currently on offer—and that the pathway concerned is independently compelling, once we rid ourselves of our temporal prejudices and look at the problem through Archimedean eyes.

Ridding ourselves of these prejudices turns out to be no easy matter, for they are deeply embedded in the ways we think about the world. Fortunately, however, it is quite easy to motivate the project, by showing that there are some very puzzling tensions in the views that most people who think about time asymmetry take for granted. In particular, there is a group of intuitions to do with the ways in which separate occurrences in the world can be connected to one another—the ways in which what happens at one place and time can depend on what happens at another place and time—which turn out to conflict with the accepted view of what physics has told us about the nature of time asymmetry in the real world. The conflict means that something has to give; as it stands, the package simply doesn't seem to hang together.

In this chapter I describe this conflict, and explain how some writers have thought—mistakenly, in my view—that it can be avoided. I also explain why there is good reason to think that it is connected to the interpretation of quantum mechanics. In effect, then, I am trying to show that there is a pathway worth exploring for two quite different reasons: first, because there is an unresolved puzzle about temporal asymmetry, and second, because it looks as though a solution to this puzzle might have very important ramifications about quantum theory.

In the following chapters we are going to tackle the exploration itself. There are two main aspects to this. On the one hand, we need to think about what we mean when we say that one event depends on another, and especially about where the temporal asymmetry involved in these notions comes from. On the other hand, we need to think about quantum mechanics, to see what kinds of problems the new pathway might be expected to solve.

Chapters 6 and 7 deal with the first of these issues. These chapters are the most philosophical in the book, and although here, as elsewhere, I have tried to avoid unnecessary technicality and to make the discussion as accessible as possible, readers with no philosophical background will find some sections tough. However, these chapters can be skimmed or skipped, at least first time through, by readers who find the ideas sketched in this chapter compelling, and want to get straight to quantum mechanics. Chapter 8 provides a survey of the peculiar puzzles of quantum mechanics, again aimed at nonspecialists, and chapter 9 explains how the new pathway makes a difference.

CONFLICTING INTUITIONS IN CONTEMPORARY PHYSICS

As we have seen, the world around us is asymmetric in time in some very striking ways. There are many processes, ranging in scale up to the expansion of the universe itself, which seem to occur with a particular temporal orientation, at least in our part of the universe. However, we have seen that since the nineteenth century, the dominant view in physics has been that these striking temporal asymmetries do not rest on any asymmetry in the laws of physics themselves. On the contrary, the laws seem essentially symmetric, in the sense that any interaction which they allow to occur with one temporal orientation is also allowed to occur with the opposite orientation (the laws showing no preference between the two). It is true that there appears to be one exception to this general principle, as I noted in chapter 1: it is the case of the decay of the neutral kaon. Even here the departure from perfect symmetry is tiny, however, and the puzzling character of the existence of this tiny exception serves to highlight the intuitive appeal of the prevailing rule. To a very large extent, then, the laws of physics seem to be blind to the direction of time—they satisfy *T-symmetry,* as we may say.

Where then does the asymmetry come from? The standard view is that it comes from what physicists call "boundary conditions." As we have seen, it is now thought that the main asymmetries stem from the fact that the universe was in a very special condition after the big bang. So the observed asymmetries are thought to stem from asymmetric boundary conditions, rather than asymmetric laws. And they are thought to be statistical in character—they are large-scale manifestations of the average behavior of huge numbers of the microscopic constituents of matter, rather than products of some asymmetry in the individual microprocesses themselves.

This view is very much the orthodoxy these days, as it has been in physics for a century or so. It is true that some physicists have challenged it, arguing that we need asymmetric laws to account for the observed temporal arrows. (I have already mentioned two writers with views of this kind: Walther Ritz [chapter 3], who argued against Einstein in 1909 that we need an asymmetric law to account for the arrow of radiation, and Roger Penrose [chapter 4], who argues that we need an asymmetric law to account for the low entropy of the big bang.) This is very much a minority view, however. For most physicists T-symmetry seems to be almost in the realm of the intuitively obvious.

However, it turns out that there is a temporally asymmetric principle, almost universally taken for granted in contemporary physics, which cannot be accommodated within this orthodox picture. The principle concerned is an extrapolation to microphysics of a familiar feature of the macroscopic world of ordinary experience: roughly, it is the principle that the properties of interacting systems are independent before they interact—though not afterwards, of course, since the interaction itself is likely to give rise to correlations. We have encountered this principle earlier in the book, of course: it is what I called PI^3, or the principle of the independence of incoming influences. It is explicitly time-asymmetric, as we saw, and has often been invoked in an attempt to explain other temporal asymmetries, such as that of thermodynamics. In general it has not been thought to conflict with T-symmetry, however. Most physicists have taken the view that like the asymmetry of thermodynamics, PI^3 is a product of asymmetric boundary conditions, rather than of the kind of lawlike asymmetry which would violate T-symmetry.

When PI^3 is extrapolated to microphysics, however, it turns out that this standard response simply doesn't work. Or rather, it turns out that if the microscopic version of PI^3 were simply a matter of boundary conditions, we would have no reason to accept it: unlike the ordinary macroscopic version and the thermodynamic asymmetry, it isn't supported on observational grounds. The upshot is that there is a real conflict between two deeply "intuitive" principles in contemporary physics: this microscopic version of PI^3, on the one hand, and T-symmetry, on the other.

In theory there are two ways we might resolve the conflict. We might leave the asymmetric principle in place, and conclude that the orthodox view described above is mistaken—that not all temporal asymmetry is statistical, and just a matter of boundary conditions. Or we might stand by T-symmetry and abandon the microscopic version of PI^3. The attractions of symmetry itself provide one argument for the latter option. Later in the chapter I want to outline another argument, which turns on the fact that the symmetric view seems to have important benefits for the interpretation of quantum mechanics. It seems to permit a very much more "classical" and less puzzling view of the microworld than quantum mechanics is usually thought to allow. For example, it seems to undermine a crucial presupposition of Bell's Theorem, which is usually taken to show that quantum mechanics involves some sort of nonlocality, or action at distance.

These connections suggest the following resolution of the conflict between T-symmetry and the microscopic version of PI^3: T-symmetry wins, not only on symmetry grounds but also because it receives strong empirical support from quantum mechanics. Quantum theory appears to confirm an hypothesis to which we might well have been led by symmetry considerations alone. And the issue sheds important new light on the physical significance of quantum mechanics itself: in effect, the microscopic version of PI³ turns out to have been obscuring from view the most attractive way to understand what quantum mechanics is telling us about the world.

PREINTERACTIVE "INNOCENCE": THE INTUITIVE ASYMMETRY

Ordinarily we take it for granted that the properties of two physical systems will be independent of one another, unless and until they interact. We assume that when two systems encounter one another for the first time, innocence precedes experience: the two systems remain blissfully unaware of each other's existence, until they first meet. This is the intuitive idea underlying PI³. We have seen that this principle has been invoked at a number of points in an attempt to explain some of the observed temporal asymmetry in the world—though usually, as I pointed out, by writers who failed to see that the real puzzle is why the corresponding principle doesn't hold in the opposite temporal direction.

Sometimes physicists and philosophers have focused on the principle itself. In 1962, for example, O. Penrose and I. C. Percival defined what they called "the law of conditional independence," which Paul Davies describes as the rule that "influences emanating from different directions in space are uncorrelated."[1] As Penrose and Percival emphasize, it is a temporally asymmetric principle. We don't expect *outgoing* influences to be uncorrelated. On the contrary, we expect systems which have interacted to have learned something from the experience, and to be correlated in some way.

As Penrose and Percival note, their principle is very closely related to what the philosopher Hans Reichenbach had earlier called "the principle of the common cause." In his book *The Direction of Time,* published posthumously in 1956, Reichenbach notes that when correlations between events at different points in space are not due to a direct causal connection, they always turn out to be associated with a joint correlation with a third event—what Reichenbach terms a *common cause*—in their common past. Reichenbach notes the temporal asymmetry of this principle, and explores the idea that it is closely connected with the issue of the direction of time. In the philosophical literature the temporal asymmetry these principles describe has come to be termed the *fork asymmetry.* The "fork" is the ∨-shaped structure of correlations, whereby two spatially separated events are correlated in virtue of their joint correlations with a third event at the vertex.[2] (As a striking example, think of the correlations between the bodily characteristics of a pair of identical twins. The correlations rest on the fact that the twins share some common history. The similarities stem from the stage where their histories overlap, and the zygote from which they both grow is the vertex of the fork.) The asymmetry of the fork asymmetry consists in the fact that forks of this kind are very common with one temporal orientation (v-shaped, or "open to the future") but rare or nonexistent with the opposite orientation (∧-shaped, or "open to the past").

At the macroscopic level of ordinary experience, the fork asymmetry appears to be connected with the asymmetry of thermodynamics. One way to see this is to imagine reversing our temporal orientation—"looking toward the past," so to speak. From this perspective we would see highly correlated *incoming* influences from distant parts of space, and these correlations would appear to be associated with the various processes in which entropy was *decreasing*. We would see the individual fragments of a wine glass hurl themselves together in precise alignment, for example, to form the unbroken glass. In other words, when we think about what would look odd about the ordinary world viewed in reverse—odd because it would seem to involve a violation of the principle that physical processes are uncorrelated before they interact—the typical cases seem to be those which exemplify the macroscopic asymmetry of thermodynamics. Or to put it in terms of our usual temporal orientation, the correlations we observe between outgoing influences are just those which result (like the thermodynamic asymmetry itself) from the rather special initial conditions which seem to prevail in the universe.

This fits in with the conclusions we drew about PI³ in chapter 3. We saw that people who try to explain the thermodynamic arrow in terms of PI³ have failed to see the nature of the puzzle about PI³ itself: why does it hold toward the future, but not toward the past? Once we see the issue in these terms, we see that the asymmetry of PI³, like the asymmetry of thermodynamics itself, rests on the fact that the universe is in a very particular condition in the direction we think of as the past.

As macroscopic principles, then, PI³ and its various manifestations appear quite compatible with the view that temporal asymmetry depends on boundary conditions, and hence with the T-symmetry of the laws of physics. However, we normally take for granted that an analogous asymmetry obtains at the microscopic level. Even concerning individual microscopic interactions, we assume that pairs of systems are correlated *after* they interact, but not *before* they do so. For example, consider something as simple and generic as the idea of an elementary particle, passing through a box or instrument of some kind. After the particle has passed through the box, we regard it as natural that its state might reflect that fact. But we are confident that this couldn't possibly be true before the particle ever reaches the box, if there has been no previous contact between the two, directly or otherwise. Without knowing anything about the nature of the particle or the contents of the box, we treat this as intuitively obvious. *Of course* the particle and the box couldn't be correlated, we think, if there has been no previous interaction between them.

As we saw earlier, we find it natural to express these intuitions in a rather anthropomorphic way, in terms of what one system may be expected to know about another. It seems intuitively obvious that interacting systems are bound to be ignorant of one another until the interaction actually occurs; at which point each system may be expected to "learn" something about the other. Let us call the microscopic case of this intuitive principle the *microscopic innocence principle* (or "*µ*Innocence," for short), to distinguish it from the ordinary or macroscopic innocence principle.

TWO KINDS OF INNOCENCE IN PHYSICS

The first thing we need to do is to show that there really are two distinct principles at stake here. The easiest way to see this is to consider the corresponding principles concerning correlations *after* interactions. In practice, we know that physical systems which have interacted in the past are often correlated in such a way that, viewed in reverse, their interaction leads to a state of lower entropy, or increased order. Viewed in reverse, the familiar world involves countless correlations of this kind; as we saw in earlier chapters, it is this that makes the reverse world look so bizarre by our familiar standards (even though it is only the actual world, viewed from a different angle).

So in the real world we find these widespread macroscopic entropy-decreasing correlations after interactions. We don't find them *before* interactions, of course, for this is just another way of saying that we don't find circumstances in which entropy decreases. This provides one sense in which real-world physical systems are uncorrelated before interactions, then: they don't display entropy-reducing correlations. This is what I mean by *macroscopic* innocence: simply the absence of entropy-decreasing correlations.

To see that this is not the only kind of preinteractive innocence to which ordinary intuition commits us, think about what we expect after interaction in a system in thermodynamic equilibrium over a long period: a gas confined to a sealed chamber for a very long time, for example. In this case the molecules of the gas do not display entropy-reducing correlations in either temporal direction; or if they do so, it is because the gas is in a low-entropy state at some point in the distant past or future, which we know nothing about, as the example has been set up. However, we still expect individual pairs of interacting molecules to exhibit postinteractive correlations. That is, we still find it natural to say that a pair of molecules which have just interacted are not independent of one another, because they have interacted. In other words, we expect a microscopic correlation of this kind, even in the absence of the kind of postinteractive correlation associated with entropy change. This shows that the two kinds of postinteractive correlation are quite different.

By symmetry, therefore, the same is true of preinteractive correlations. We can make sense of a kind of preinteractive correlation which would not require a violation of the second law, namely the temporal inverse of the kind of microscopic postinteractive correlation we do expect to find in the case just described. μInnocence is the assumption that there are no such microscopic preinteractive correlations. The example illustrates the principle's intuitive plausibility: although we find it natural to say that particles which have interacted in the recent past are not independent, we regard it as obvious that this is an asymmetric matter. Particles which interact in the near future seem no less clearly independent—innocent of one another's very existence—than those which never meet at all.

IS μINNOCENCE OBSERVABLE?

To the extent that the asymmetry of μInnocence has been noticed in physics, it has been thought to have the same status as that of the macroscopic ("no entropy-reducing correlations") version of the innocence principle: both versions are regarded as products of asymmetric boundary conditions, rather than as asymmetric laws. (Since the distinction between the two versions has never been clearly drawn, so far as I know, it is hardly surprising that they have been treated in the same way.)

In fact, however, there is an important difference between the two cases. In the macroscopic case, countless manifestations of the second law of thermodynamics provide direct evidence for the existence of an asymmetry concerning entropy-reducing correlations. In effect, we can simply observe that these correlations are common with one temporal orientation, and yet very rare with the other. In other words, we have very good empirical reasons to accept the existence of the asymmetry, independently of any claim about boundary conditions required to explain it.

In the case of μInnocence, however, there seems to be no *observed* asymmetry to be explained. It is not that we *observe* that the incoming particle is not correlated with the contents of the box through which it is about to pass. Rather, we rely on the asymmetric principle that interaction *produces* correlations only in one temporal direction—"toward the future," and not "toward the past."

Without an observational basis for μInnocence, however, the only explanation for the almost unchallenged status it enjoys in physics would be that it is regarded as a fundamental principle in its own right. This would mean that contemporary physics does take it for granted that there is an asymmetry in the boundary conditions of the kind required by μInnocence—but only because it takes for granted (asymmetrically, and in violation of T-symmetry) that the dynamics of interactions produces correlations in the future but not the past, and not because it has any direct evidence for an asymmetry in the boundary conditions themselves. In these circumstances, to argue that the asymmetry of μInnocence is merely a matter of boundary conditions would be to undermine one's own reasons for accepting that there is any such asymmetry in the first place.

Two objections tend to be raised at this point. The first claims that μInnocence does not need observational support, because it can be ruled out on the grounds that its failure would require "miraculous" correlations between systems which are about to interact. By now it should be clear what is wrong with this kind of objection, however. If it amounts to anything more than a bald assertion of μInnocence itself, then it is an attempt to appeal to statistical or option-counting considerations. As we have seen many times in the three previous chapters, however, these considerations are symmetric in themselves. Postinteractive correlations don't *seem* "miraculous," but this is because we already take for granted the very asymmetric principle at issue, namely that interaction produces correlations "to the future" but not "to the past."

The second objection is more interesting. Physicists often claim that μInnocence is observable, albeit indirectly. They point to a range of cases in which the assumption of μInnocence seems to enable us to explain observed asymmetries. (Penrose and Percival themselves describe several examples of this kind, in support of their principle of conditional independence.) A typical case involves the scattering of intersecting beams of particles. Consider what happens when two narrow jets of water intersect, for example: water scatters in many different directions. The suggestion is that this can be explained if we assume that the individual pairs of colliding molecules—one from each stream—are uncorrelated with one another. If μInnocence can be used in this way, doesn't this amount to indirect observational evidence in favor of the principle?

My view is that these examples rely on a confusion between μInnocence and the macroscopic innocence principle, and a failure to see that what the explanations in question actually require is only the latter principle. For example, let's think about the water jet case more carefully. If μInnocence is really crucial here, then it ought to be the fact that μInnocence doesn't hold for *post*-interactive correlations that explains why we don't see scattering as we look toward the past. (In this direction, we see the water forming itself into two narrow jets, and converging on two small nozzles.) In this direction, however, microscopic postinteractive correlations between pairs of water molecules are not enough to explain what goes on: we need widespread patterns of entropy-reducing correlations, in order to explain the fact that the water takes on such an orderly form. By symmetry, then, it is the absence of entropy-reducing correlations which we need to explain what we see toward the future; that is, to explain why the water scatters, rather than, say, forming into new jets. The observed asymmetry turns on the asymmetry at the level of widespread or macroscopic entropy-reducing correlations, rather than on an asymmetry at the level of the μInnocence principle. I think the same is true of all the examples which physicists take to confirm the lack of preinteractive correlations.

In other words, there seems to be no observational evidence for any temporal asymmetry in pre- and postinteractive correlations, other than the kind of widespread macroscopic correlation associated with the thermodynamic arrow. In particular, there seems to be no observational evidence for μInnocence. This means that μInnocence cannot be reconciled with T-symmetry on the basis that it simply involves an asymmetry in boundary conditions. As it currently operates in physics, μInnocence is a freestanding principle. In effect, we accept that there is an asymmetry in boundary conditions because we accept this principle, and not the other way around. This has been obscured by a failure to distinguish between the kind of correlation associated with increase or decrease of entropy, and the kind we take to be produced by microscopic interactions. Observation reveals an asymmetry in correlations of the former kind, but μInnocence requires an asymmetry in correlations of the latter kind.

SYMMETRY OR INNOCENCE?

This means that there really is a conflict in the intuitive picture of the world with which contemporary physics operates. Given that it doesn't have an observational basis, our intuitive commitment to μInnocence is incompatible with T-symmetry.

How should we address this conflict? If observational factors were entirely irrelevant, then T-symmetry would have a strong case, on symmetry grounds alone. Symmetric solutions are generally the default options in science, from which departures need to be justified. In the present case, this suggests that in the absence of an explicit reason to think otherwise, we should assume that interaction in the future is no less (and of course, no more) a source of correlations than interaction in the past. In other words, we should take the view that an interaction in the future is just as much a reason to think that two systems are not independent of one another as is an interaction in the past. True, this is deeply counterintuitive. However, my point has been that once we distinguish the asymmetry of μInnocence from the macroscopic asymmetry associated with the second law of thermodynamics, the deep intuition concerned turns out to be not only groundless, but also in conflict with accepted principles of symmetry in microphysics.

However, it is not clear that the issue does need to be decided simply in these terms. To say that μInnocence does not *currently* rest on observational considerations is not to say that observational evidence is necessarily irrelevant to the issue as to whether it holds. But what kind of empirical evidence would make a difference, one way or the other? In particular, what should we expect the world to be like if there were no asymmetry of μInnocence—if, at the microscopic level, interacting systems were correlated in the same way both before and after their interaction? Remarkably enough, it seems that the right answer to this question may be this one: *We should expect the world to be the kind of world we find in quantum mechanics.* Quantum mechanics seems to describe the kind of world we ought to have expected, had we considered rejecting μInnocence in the first place.

μINNOCENCE AND QUANTUM MECHANICS

It is often said that quantum mechanics is both the most successful theory physics has ever produced, and the most puzzling. The quantum world is strange and "nonclassical" in radical and disturbing ways. Exactly how it is strange is a matter for debate: there are many competing accounts of what quantum theory actually does tell us about the world, each giving its own view of what is nonclassical about quantum theory. To a large extent, however, the disagreements between these different interpretations come down to a choice between evils. As we shall see in chapter 8, problems avoided in one place tend to break out in another.

All the same, there are certain key results of quantum theory which almost everybody in these debates takes to be crucial, and to embody the nonclassical "strangeness" of the quantum world. Perhaps the most important of all is the result known as Bell's Theorem, established by the Irish physicist John Bell (1930–1990). Bell's result has been taken to show that the quantum world involves some kind of "nonlocality," or action-at-a-distance. Action-at-a-distance has usually been regarded with suspicion in physics, and in the contemporary context is doubtfully compatible with Einstein's theory of special relativity. Bell's Theorem has thus been taken to reveal a tension between two of the most important pillars of modern physics, quantum theory and special relativity. While the nature of quantum nonlocality and the extent of its conflict with special relativity have been a matter for much debate, the consensus has been that quantum mechanics is committed to it in some form.

As Bell and others have pointed out, however, Bell's Theorem depends on the assumption that quantum systems are not correlated with the settings of measurement devices, prior to their interaction. Thanks to μInnocence, this assumption has normally seemed uncontentious. Bell himself considered relaxing it, but even he tended to think about this possibility in a way which doesn't conflict with μInnocence. (His suggestion, which he called "superdeterminism," was that the required correlation might be established by an additional common cause in the past, not simply in virtue of the existing interaction in the future; more on this in chapter 9.) The upshot is that without μInnocence, there seems to be no reason to think that quantum mechanics commits us to nonlocality. It turns out that this applies not simply to Bell's Theorem, but also to other more recent arguments for nonlocality; these too depend on μInnocence, in the form of the assumption that the states of incoming particles are not correlated with the settings of instruments they are yet to encounter.

Nonlocality isn't the only nonclassical consequence thought to flow from quantum mechanics, of course. Since the early years of quantum theory, many physicists have been convinced that quantum measurements do not simply reveal a preexisting classical reality. In Bohr's "Copenhagen Interpretation," which remains very influential, the view was that reality is somehow "indeterminate" until a measurement is made—measurement was said to force reality to take on a definite condition, where none existed before. Later, a range of mathematical results—the so-called no hidden variable theorems— seemed to establish that no system of preexisting properties could reproduce the predictions of quantum mechanics, at least in certain cases. (Again, we will look at these arguments in a little more detail in chapter 8.)

These interpretations and results also take for granted μInnocence, however. Otherwise, they would not be entitled to assume that the *preexisting* reality could not depend on the nature of a later measurement. In place of Bohr's indeterminate reality, one might have postulated a reality which, while fully determinate before a measurement is made, is partly constrained by the nature of that measurement. In the case of the no hidden variable theorems, similarly, μInnocence serves to justify the assumption that a single hidden state should be required to reproduce the quantum predictions for any possible next measurement. Without μInnocence, we would expect to find different hidden states in otherwise similar systems which were going to be involved in different interactions in the future. After all, this seems to be just what it would mean for systems to be correlated in virtue of future interactions, as well as past ones. So if the hidden state is allowed to vary with the nature of the upcoming measurement, the problem of finding a hidden variable theory is relatively trivial.

There are some hidden variable theories already in quantum theory, despite the no hidden variable results. The best known is that of the late David Bohm, a London-based American physicist who worked outside the mainstream in quantum theory for many years, and whose ideas are now the focus of renewed attention.[3] Bohm's theory escapes the no hidden variable theorems by allowing measurement to have an instantaneous effect on the hidden variables. Again, however, it is μInnocence which underpins the assumption that the effect must be an instantaneous one, rather than one which occurs in advance.

μInnocence thus plays a crucial role in the arguments which are taken to show that quantum mechanics has puzzling nonclassical consequences. In other words, it seems to be only if we add μInnocence to quantum mechanics that we get the problematic consequences. We might symbolize the logical relationship like this:

$$QM + \mu\text{Innocence} \Rightarrow \text{Nonlocality, Indeterminacy, ...}$$

On the right-hand side we have a list of the problematic consequences of quantum mechanics. The contents of the list vary a little with one's favored interpretation of quantum theory, but virtually everyone agrees that there are some such consequences.

To understand the significance of these connections, try to imagine how things would have looked if we had considered abandoning μInnocence on symmetry grounds, *before the development of quantum mechanics*. (This idea doesn't seem particularly far-fetched, incidentally, for T-symmetry receives its main impetus from classical sources—especially the statistical treatment of irreversibility in thermodynamics.) Quantum mechanics would then have seemed to provide a dramatic *confirmation* of the hypothesis that μInnocence fails: for, given quantum mechanics, the assumption that μInnocence does not fail turns out to imply such apparent absurdities such as nonlocality and indeterminacy. Against this imagined background, for example, recent experimental confirmations of the predictions of quantum mechanics involved in Bell's Theorem would have seemed to provide *observational* data for which the only reasonable explanation is that μInnocence does fail, as already predicted on symmetry grounds.

From a contemporary standpoint it is very difficult to see the issue in these terms, of course. We are so used to talk of nonlocality and indeterminacy in quantum mechanics that they no longer seem entirely absurd. And of course we are still so strongly committed to *μ*Innocence that it is hard to see that rejecting it could provide a more plausible way to understand the lessons of quantum mechanics. But I think it is worth making the effort to challenge our preconceptions. I have argued that we have reason to doubt *μ*Innocence on purely classical grounds—simply for symmetry reasons, in effect, once we appreciate that we have no empirical reason to question T-symmetry. With this new conception of the proper form of a *classical* microphysics, it seems unwise to continue to insist that quantum mechanics is a radically *nonclassical* theory, in what have become the accepted ways.

*μ*INNOCENCE AND BACKWARD CAUSATION

Thus the hypothesis that *μ*Innocence fails seems to throw open the conventional debate about quantum mechanics in a rather appealing way—it suggests that quantum mechanics might be a very much less nonclassical theory than almost everybody has assumed. But is the hypothesis really one to be taken seriously? The appeal to symmetry notwithstanding, many will feel that there is something fundamentally absurd about the suggestion that *μ*Innocence might fail; that physical systems might "know" something about one another, before they ever interact.

A comprehensive response to these doubts will require a much better understanding of the significance of notions such as physical dependence—that is, the notion we use when we say that the state of one system is dependent on, or independent of, the state of another. The next two chapters will be concerned with some of the delicate philosophical work of understanding these superficially familiar notions. In order to give a flavor of the likely consequences of abandoning *μ*Innocence, however, I want to finish here by mentioning one surprising consequence which turns out to be very much less objectionable than it seems at first sight. In the process, I think, it provides further ammunition for the claim that quantum theory provides precisely the kind of picture of the microworld we should have expected, if we had accepted in advance that T-symmetry requires us to abandon *μ*Innocence.

The consequence concerned would show up in a case in which we had influence over one member of a pair of interacting systems—over the setting of the contents of the box in our earlier example, say, before its encounter with the incoming particle. If the state of the incoming particle were correlated with that of the box, then in controlling the box (or its contents) we would be able to control the particle. This would not be action at a distance—the correlation would be conducted continuously, via the interacting worldlines of the two systems involved—but it would seem to amount to a kind of "backward" or "advanced" causation. This consequence might well seem absurd, and potentially paradoxical.

The usual objection to advanced causation involves what philosophers call the *bilking argument.* We will look at this in more detail in chapter 7, but the essential idea is that in order to disprove a claim of advanced causation, we need only arrange things so that the claimed later cause occurs when the claimed earlier effect has been observed not to occur, and vice versa. In other words, we arrange things so that there would be no noncontradictory outcome possible, given the claimed causal links. It might be thought that an argument of this kind will be sufficient to defend μInnocence. Any claimed preinteractive correlation looks liable to be disproved in this fashion—we need only ensure that the properties of one system affect those of the other, in such a way as to conflict with the claimed correlation.

But is the bilking argument effective in the kind of case we are considering? Consider our imaginary example. In order to set up the kind of experiment just outlined, we would need to observe the relevant state of the incoming particle before it reaches the box, so as to arrange the contents of the box in such a way as to defeat the claim that the incoming particle is correlated with the state of the box. But how would we set about making such an observation? Presumably we would have to place a second box, or some other measuring device, in the path of the particle, before it reaches the original box. But if we are entertaining the hypothesis that μInnocence fails, we have two reasons to dispute the relevance of the information yielded by this measurement procedure.

First, if μInnocence is in doubt then we are not entitled to assume that the state revealed by this measurement is the state the particle *would have had,* had the measurement not been made. After all, if measurements can affect the earlier states of the systems measured, then what measurement reveals is not in general what would have been there, in the absence of the measurement in question. Even if we found that the correlation required for backward causation failed in the presence of the measuring device (the second box), in other words, we would not be entitled to conclude that it would have failed in its absence.

Second, and more important, what the failure of μInnocence requires is that there be a correlation between the box setting and the state of the incoming particle *from the time of that particle's last interaction with something else.* For think of the usual case: how long do we expect a correlation established by interaction to survive? Not beyond the time at which the system in question interacts with something else, a process which may destroy the initial correlation. In the case we are considering, then, the effect of interposing a second measuring device in the particle's track will be to ensure that the correlation is confined to the interval between this measurement and the main interaction. The presence of the measurement ensures that the advanced effect is more limited in extent than it would be otherwise, and again the required contradiction slips out of reach.

These objections could be evaded if it were possible to observe the state of the incoming particle without disturbing it in any way—if the presence of the measuring device made no difference to the object system, in effect. Classical physics is often said to have taken for granted that perfectly nonintrusive measurements of this kind are possible, in principle, or at least approachable without limit. If this view was ever assumed by classical physics, however, it was decisively overturned by quantum mechanics. One of the few things that all commentators agree on about quantum mechanics is that it shows that we cannot separate the processes of measurement from the behavior of the systems observed. This provides a further respect in which quantum mechanics is the kind of microphysical theory we might have expected, if we had questioned μInnocence on symmetry grounds, within the classical framework. The bilking argument suggests that classical nonintrusive measurement is incompatible with the kind of symmetry required if we abandon μInnocence. Turning this around, then, it seems that symmetry considerations alone might have led us to predict the demise of classical measurement, on the grounds that it is incompatible with microscopic T-symmetry.

THE NEXT STEP

Summing up, our puzzle stems from the conflict between two very plausible principles. On the one hand, the physics of the past 100 years seems to give us good reason to accept that the laws governing the microscopic constituents of matter are insensitive to the distinction between past and future. On the other hand, we normally take for granted that on the small scale, as on the large, interacting systems do not become acquainted before they actually meet. It seems that one of these intuitions has to go, but which one? Can physics help us out here? In particular, could there be empirical evidence one way or other?

I have suggested that there is already strong evidence in favor of the symmetric alternative, although evidence of an indirect kind. From a classical standpoint, quantum mechanics itself is naturally taken to provide such evidence, on the grounds that if combined with the principle of μInnocence, it leads to such conceptual horrors as nonlocality and indeterminacy. From a contemporary standpoint, however, these ideas have lost their capacity to shock. Familiarity has bred contentment in physics, and the reductio has lost its absurdum. Regaining a classical perspective will not be an easy step, nor one to be attempted lightly, but it does seem worth considering. In abandoning a habit of thought which already conflicts with well-established principles of symmetry, we stand to free quantum mechanics of metaphysical commitments which once seemed intolerable in physics, and might well do so again. The crucial point is that this isn't simply a matter of trading one set of implausibilities for another (the kind of trade-off which always characterizes the choice between rival interpretations of quantum mechanics). The choice between T-symmetry and μInnocence needs to be made in any case. As long as we favor symmetry, the advantages for quantum mechanics come at no extra cost.

For the moment this is all very promissory. In effect, I have called attention to the outlines of the doorway, but have done little to show that it is even in working order, let alone that it actually leads anywhere. The next step is to try to show that the claims of the advertisement stand up to scrutiny, that the promise of this approach is not illusory, and hence that cautious and skeptical readers do not risk their reputations if they step across the threshold and take a look for themselves.

There are two parts to this next step. One part is a matter of saying more about quantum mechanics, to fill out the claim that abandoning μInnocence pays big dividends concerning the interpretation of quantum theory. The other part is more philosophical. We have seen μInnocence is so closely associated with our intuitions about causation: without μInnocence, for one thing, it seems possible for our present choices to affect the past. So in order to get a clear view of μInnocence, and especially of the proposal to give it up, we need to tackle the notion of causation itself.

One relevant factor is that causation itself involves a striking temporal arrow. In general, causes always seem to occur *before* their effects. Like the physical arrows we looked at earlier, this causal arrow is rather puzzling, especially in light of the apparent temporal symmetry of the laws of physics. Why should causation show this strong temporal bias? And how is the fact that it does so related to μInnocence?

Given that abandoning μInnocence seems to lead to backward causation, a natural thought is that the asymmetry of μInnocence and the arrow of causation amount to much the same thing. If so, then in arguing against μInnocence, I would also be arguing against the standard view of causation; maintaining that here, too, the view we take for granted turns out to be unsupported by physics.

This strategy would have the advantage that it hitched the campaign against μInnocence to a much larger wagon. By the same token, however, it would raise the stakes by several orders of magnitude, in seeking to over-turn not simply the intuitions of physicists in microphysics, but all of our ordinary intuitions about cause and effect. I think this strategy is the wrong one, however. In the end, μInnocence and the arrow of causation turn out to be separate issues. The campaign against μInnocence does not involve a general assault on ordinary notions of cause and effect. In order to see that this is the case, however, we need to understand the source of the asymmetry in the ordinary notions.

In effect, then, we need to think about the familiar arrow of causation, before we can think clearly about the idea of giving up μInnocence. In the next two chapters I argue for a particular solution to the puzzle of the causal arrow. Roughly, it is that the asymmetry and temporal orientation of causa-tion rests on a subjective element in the ordinary notion of causation. The directionality comes from the asymmetry of our perspective, on this view, and not from any asymmetric ingredient in the world.

The idea that the asymmetry of causation is subjective has often ap-pealed to physicists, struck by the thought that the underlying T-symmetry of physics seems to leave no place for asymmetric causation. Ironically, how-ever, this has made it harder rather than easier to see the possibility of a backward causation approach to quantum mechanics, based on abandoning μInnocence. After all, if the direction of causation is subjective, or perspec-tival, how could there be anything in the claim that quantum mechanics shows that there is backward as well as forward causation?

In fact, there is no contradiction here. Abandoning μInnocence turns out to be a thoroughly objective proposal, though one which introduces what is naturally described as backward causation, *from the ordinary asymmetric perspective.* In order to get a clear view of the objective significance of aban-doning μInnocence, however, it is important to disentangle the subjective and objective elements.

In the next two chapters I develop this perspectival approach to the prob-lem of the direction of causation, and show how it leaves room for the proposal to abandon μInnocence. In philosophical terms, these chapters are probably the densest in the book. Nonphilosophical readers should feel free to skim to the end of chapter 7, where the discussion ties in once more with the themes of this chapter.

6

←——————→

In Search of the Third Arrow

Our next topic relies less on the discoveries of physics than on intuitions about the world which all of us share. In order to show that these folk intuitions are not divorced from physics, however, I would like to begin with a simple physical example. I would like you to imagine that you are wearing your sunglasses one cloudless summer evening—it is one of those neighborhoods where people wear their sunglasses at night, drink serious coffee, and dress in black. As you stare at what you can see of the night sky, photons from distant stars pass through the polarizing lenses of your sunglasses, through the focusing lenses of your eyeball, and are detected on your retina. Imagine now that some of these photons have encountered polarizing lenses before. Early in their journey from a distant star, they happened to pass through some sunglasses carelessly discarded by an alien astronaut. Consider the state of these photons as they near the end of their journey. The intuition I ask you to share is this one. The state of the photons does not depend on the fact that they are going to pass through a polarizing lens in the very near future, but it may depend on the fact that they have passed through such a lens in the distant past. It may depend on the past, but it doesn't depend on the future.

At this stage what I would like you to acknowledge is that this intuition is natural and compelling. You may be troubled by its imprecision, and feel the urge to try to formulate it more precisely. If so, bear with me. For the moment what interests us is that the intuition reveals a very stark temporal asymmetry in our view of the world. We take it for granted that things depend on what happens at earlier times but not on what happens at later times. Once again, this is an asymmetry so obvious that normally we simply don't see it. And when we do notice it, what makes it puzzling is again that there seems to be no basis for such an asymmetry in the physical phenomena themselves. Why should dependence on the future polarizers be any more problematic than dependence on the earlier polarizers? What is there in the photon's world to sustain such a difference, particularly if—as we are assuming—there is no objective sense in which the photon travels from the star to the Earth, rather than vice versa? Where does this asymmetry of dependence come from?

This seems to be a version of the puzzle we encountered in chapter 5, of course. After all, the case of photons and sunglasses is just a less anonymous version of our example involving boxes and incoming particles. However, it is worth noting that the asymmetry of dependence also seems to be a feature of much more commonplace examples. Consider a simple collision between two billiard balls, for example. After the collision, the motion of the red ball depends on the fact that it was struck by the white ball; *after* the collision, certainly, but not *before* it. If this, too, is a matter of μInnocence, then μInnocence hardly seems a controversial principle of microphysics.

In this chapter and the next our project is to gain a better understanding of the kind of intuitions to which these examples appeal. Eventually, I want to tease apart μInnocence and what turns out to be a noncontroversial kind of asymmetry of dependence, and to show that there is an important difference between the photons and the billiard balls: only in the case of our intuitions about the photons is μInnocence at work.

Our immediate focus will be on the asymmetry of dependence, and on two closely related temporal asymmetries. We thus have three targets in all:

- *The temporal asymmetry of dependence.* Events often depend on what happens at *earlier* times, but never (or almost never)[1] on what happens at *later* times.

- *The temporal asymmetry of causation.* Effects occur *after* but never (or almost never) *before* their causes.

- *The temporal asymmetry of agency.* Human actions influence *later* events but not *earlier* events.

The last of these notions is more anthropocentric than the other two, and might seem marginal or derivative, if our interest is in the time asymmetry of the objective world. However, we'll see that one of the crucial issues in this area concerns the extent to which the asymmetries of dependence and causation are themselves "fully objective." I am going to argue that here, even more than elsewhere in the study of time asymmetry, it is crucial to understand and disentangle the subjective component in our view of the world. In a sense I'll explain as we go along, the asymmetry of agency thus comes to play a crucial role in our understanding of causation and dependence.

The idea that causation is at least partly a subjective notion is common among physicists, who are inclined to see it as a reason for regarding the temporal asymmetry of causation as physically uninteresting. A common view is that the only physically respectable notion of causation is one which we know to be temporally symmetric, namely the notion of what may be deduced from what in accordance with deterministic laws. For example, in a recent paper Stephen Hawking describes his early encounter with Hans Reichenbach's *The Direction of Time*:

> [W]hen I eventually got hold of the book, I was very disappointed. It was rather obscure, and the logic seemed to be circular. It laid great stress on causation, in distinguishing the forward direction of time from the backward direction. But in physics we believe that there are laws that determine the evolution of the universe uniquely. So if state A evolved into state B, one could say that A caused B. But one could equally well look at it in the other direction of time, and say that B caused A. So causality does not define a direction of time.[2]

However, physicists who dismiss the ordinary asymmetric notions of causality in this way usually fail to notice how these very notions constrain their ideas of what counts as an acceptable physical theory. The simple photon example provides an illustration. We find it natural that the state of the photon might depend on, or be correlated with, the state of the distant polarizer through which it passed as it left the region of the distant star. On the other hand we find it almost absurd to suggest that it might be similarly correlated with the state of the local polarizer (the lens of our sunglasses, through which it hasn't yet passed). On the face of it, the view that these intuitions are anthropocentric in origin seems committed to saying that physics should treat both cases in exactly the same way. As we shall see, however, almost nobody in physics thinks of that as a natural way to proceed.

Another illustration of this point is provided by contemporary quantum mechanics. In chapter 8 we shall see that as standardly interpreted, quantum theory embodies what seems to be a concrete representation of the fact that the state of the photon depends on its past but not its future. On the standard model, the actual quantum mechanical state of the photon as it approaches the Earth reflects the fact that it *has passed* the distant earlier polarizer, but not the fact that it *will pass* the local future polarizer. This seems flatly incompatible with the view that the asymmetry of dependence is merely subjective in origin. Once again, then, physics does not practice what many physicists preach in this respect.

I emphasize that I am not endorsing the standard practice. On the contrary, I think it is a very unsatisfactory feature of the standard model of quantum theory that it does admit this striking temporal asymmetry, so out of keeping with the temporal symmetry of (almost all of) the rest of physics. Indeed, as I explained in chapter 5, I want to argue that the most profound difficulties in the interpretation of quantum mechanics simply dissolve if quantum mechanical systems are allowed to "know" about their future in certain ways. In my view this avenue to a satisfactory interpretation of quantum mechanics tends to be overlooked precisely because the causal intuitions we are dealing with remain absolutely robust in the minds of those who contemplate the quantum mysteries. So my claim is much more that physicists should begin to practice what they preach concerning the subjectivity of causal asymmetry, than that they should change their tune. However, I think that to be in a position to resist the pull of the familiar intuitions, it is necessary to "know one's enemy"—that is, to understand how these intuitions arise, and why they are so powerful. So we cannot avoid the careful work of this chapter and the next.

One further illustration before we begin. Consider John Wheeler, discussing what is known as the delayed choice two-slit experiment. This is a version of the standard two-slit experiment, famous in quantum theory, which has long been thought to reveal the essential wave-particle duality of the quantum world. The new feature of the delayed choice case, introduced by Wheeler, is that the choice between the one-slit and two-slit versions of the experiment is not made until after the particles concerned have passed through the apparatus. Wheeler points out that this makes no difference to the quantum predictions, and asks:

> Does this result mean that present choice influences past dynamics, *in contravention of every principle of causality?* Or does it mean, calculate pedantically and don't ask questions? Neither; the lesson presents itself rather as this, that the past has no existence except as it is recorded in the present.[3]

But what are these principles of causality to which Wheeler refers? The example illustrates two things: first, that whatever they are, they don't appear to be built into the physics itself, being rather some sort of metaphysical constraint on acceptable interpretations of the physics; and second, that they are so intuitively powerful that theorists such as Wheeler prefer to sacrifice the observer-independent existence of the past than to give them up. Indeed, Wheeler is somewhat exceptional in that he actually notices the option of violating principles of causality, as he puts it. If a metaphysical principle can play such a powerful role in constraining the interpretation of quantum mechanics—indeed, if it can play this role despite the widespread view that the matters it deals with are "merely subjective"—then surely it is worthy of the attention of physics.

The above remarks were directed at those who might agree with me that the temporal asymmetry of causation is partly subjective in origin, but who might therefore be inclined to dismiss the concerns of this chapter and the next as of little relevance to physics. For the moment, however, my main opponents are those who disagree at the first step: the many philosophers (and no doubt many physicists) who do not find it at all obvious that there is something subjective and perspectival about the temporal asymmetries of causation and dependence. For the moment, then, the main task is to present the case for this view.

This chapter focuses on the asymmetry of causation. After clarifying the issues a little, I am going to criticize the leading rival to the view I want to defend. This rival view attempts to account for the temporal asymmetry of causation in terms of a physical asymmetry in the world—typically, as we shall see, an asymmetry closely related to some of those we discussed in earlier chapters. My main objection to this approach draws on the conclusions of those chapters: I argue that the physical asymmetries concerned have the wrong character and the wrong distribution to provide the basis for the asymmetry of causation. I shall also indicate why several alternative approaches seem to me to fail to provide what we need from an account of causal asymmetry, and close by outlining what seems to me a more promising approach, namely the perspectival view I foreshadowed earlier.

Chapter 7 explores and defends this perspectival account. Returning to the simple photon example, we shall see that according to a plausible account of the meaning of dependence claims, at least one aspect of their time asymmetry rests on an asymmetry in the conventions governing their use. This suggests that there is an important sense in which the asymmetry lies in our linguistic practices, rather than in any objective fact in the world. The natural objection to this proposal is that it cannot account for the apparent objectivity of the asymmetry of dependence. It seems to make it merely a conventional matter that we can affect the future but not the past, for example. Responding to this objection will help us to clarify the confusing interplay between subjective and objective factors in this area. Eventually, it will reveal that even when all subjective factors are set aside, there is still an objective asymmetry revealed in our intuitions concerning the photon case; in effect, it confirms that in assuming μInnocence, we are assuming an objective temporal asymmetry in the physical world.

CAUSAL ASYMMETRY: THE NATURE OF THE PROBLEM

We are interested in why effects don't occur *before* their causes—in other words, in why the cause-effect relation always seems to be aligned "past to future," and not "future to past." (For simplicity let us ignore for the time being the possibility of rare cases of backward causation, and also possible cases of simultaneous causation.) Notice that this issue presupposes that the cause-effect relation is itself asymmetric—that is, that causes and effects can be distinguished in some way. After all, if we had a compass arrow which looked the same at both ends, it wouldn't make sense to ask whether it was pointing north rather than south. So the problem of the temporal asymmetry of causation splits into two subproblems: one task of a theory of causation is to explain the difference between causes and effects, to reveal the true point of the internal "arrow" of causation, so to speak; another is to explain why this internal arrow of causation is so well aligned with the arrow of time.

Once we notice that the problem has two parts to it, however, it is tempting to kill two birds with one fiat, so to speak, by saying that the two arrows are related by definition, or linguistic convention. According to this view the cause-effect relation doesn't involve any *intrinsic* asymmetry in the world. Rather we simply use the different terms *cause* and *effect* to distinguish the earlier and later members of a pair of events which are related in some symmetric way. The phrase "is a cause of" is shorthand for something like "is earlier than and causally related to," where "causally related to" refers to a symmetric relationship. Famously, the Scottish philosopher David Hume (1711–1776) was one writer who took this course. According to the orthodox interpretation of Hume's view, causation is simply a matter of constant conjunction—itself a temporally symmetric notion—and we use the term "cause" to mark the earlier and "effect" to mark the later of a pair of events which are related in this way.

By way of comparison, think of the way in which the acts of a play are normally numbered. The act performed first is called "Act 1," and so on. Obviously this is just a convention. Plays could be numbered like countdowns, for example, so that Act 1, Scene 1 would be the end of the play. Similarly, the Humean view is that the terms "cause" and "effect" are just conventional labels, which we use to mark the earlier and the later members of a pair of suitably related events.

Philosophers have often pointed out that there seems to be a heavy price for the convenience of this conventionalist approach, however. On the one hand, it seems to make it a merely linguistic matter—a matter of what we mean by "affect," in effect—that we can affect the future but not the past. This seems too weak: our intuition is that there is more to the difference than that. On the other hand, it also seems to make it a logical impossibility—again, a matter of the meaning of the terms—that there is no simultaneous causation and backward causation, and this seems to many philosophers to be too strong. Perhaps there is no simultaneous or backward causation, these philosophers think, but surely the idea of these things is not self-contradictory, as it would be if "causes precede their effects" were an analytic truth, like "a spouse is a marital partner." Some philosophers also object that the Humean strategy precludes the project, attractive to many, of explicating temporal order in terms of causal order.[4]

In chapter 7 I am going to argue that a more sophisticated version of the conventionalist approach—a version a good deal less arbitrary than Hume's—can meet most of these objections. (It doesn't meet the last one, but in my view this is a good thing. Since we have good reason to think that there is no intrinsic direction to time itself, we shouldn't look to causation to provide one!) For the moment, however, I want to consider the alternatives.

A THIRD ARROW?

In rejecting the Humean view that the asymmetry of the causal relation is merely a conventional image of earlier-later ordering, contemporary writers have often suggested that it rests on some intermediate asymmetry—on some temporally asymmetric feature of the world which is itself aligned (typically, if not invariably) with the temporal ordering. This is what I shall call the *third arrow strategy*. Most commonly, advocates of this approach argue that when we analyze causation we find that we can characterize the essential relationship between a cause and an effect in physical terms—and that the physical relationship concerned turns out to hold asymmetrically in the real world. Often the latter fact is held to be one that just happens to hold in the world as we know it. It is held to be a de facto asymmetry, rather than one required by the laws of physics, for example. So we end up with a view something like that shown in Figure 6.1. In assessing this approach to causal asymmetry, then, the essential question is whether there is actually an intermediate temporal asymmetry in the world of the required sort. By examining what seems to be the only plausible candidate, I want to argue that there is not.

THE FORK ASYMMETRY

This solitary candidate is an asymmetry we have already encountered. In the early chapters of the book we often came across the idea that outgoing processes from a common center tend to be correlated with one another, whereas incoming processes meeting at a common center always seem to be uncorrelated, or statistically independent. The latter point was essentially our PI^3, while the former is well exemplified by examples of waves on ponds and the like. A strong tradition in modern philosophy, stemming largely

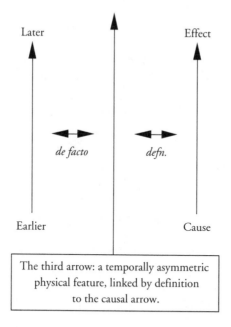

Figure 6.1. The third arrow strategy.

from the work of Reichenbach in the 1950s, attempts to link this apparent temporal asymmetry to that of causation. As I explained in chapter 5, the temporal asymmetry of patterns of correlation is called the *fork asymmetry*. In my terminology, the central theme of this tradition is that the fork asymmetry provides the missing third arrow which underlies causal asymmetry.

If the fork asymmetry is to play this role, we must be careful to characterize it without using the very causal notions we hope to cash out in terms of it—for otherwise, of course, we don't have an independent asymmetry, in terms of which to account for that of causation. Keeping this caution in mind, we find that there are two main ingredients to what is described as the fork asymmetry. Loosely speaking, the first ingredient is the principle that *outgoing* processes from a common center tend to be correlated, whereas *incoming* processes do not.[5] When a "fork" of processes has its "prongs" pointing to the *future,* these prongs tend to be correlated; not so for forks of processes whose prongs point to the *past.* (The italicized terms assume a temporal orientation, of course, but no immediate harm comes of this.) In other words, this first ingredient tells us that *if* certain conditions obtain—if there is a suitable central event in the common past—*then* we tend to find correlations between spatially separated events which are not themselves directly connected.[6] The second ingredient is the converse principle: it tells us that we *only* find such correlations between spatially separated events under these conditions—in other words, that correlations of the specified kind *only* arise in association with outgoing processes from an *earlier* common center.

In precausal terms, then, the fork asymmetry comprises a striking feature of the correlational structure of the world. It amounts to the fact that we find v-shaped correlational forks of the kind which are open to the future but not the kind which are open to the past, and that all nonlocal correlations of the appropriate sort belong to such a structure.

There are two main ways to try to link the fork asymmetry to that of causation. One approach is to say that it is *definitive* of causes (as against effects) that combining causal factors are independent, and then to use the fork asymmetry to explain why causes typically occur before their effects. Another is to take the role of explaining nonlocal correlations to be definitive of causes, and then to say that the fork asymmetry tells us that in practice events which play this coordinating role lie in the past of the events whose correlation they explain. (These approaches are really very close to one another. Roughly speaking, the first simply looks at the prongs of the future-directed forks we actually find in the world and says, "These are effects", while the second looks at the vertices and says, "These are causes.")

Once the connection between the fork asymmetry and causation has been drawn, it may be convenient to formulate the fork asymmetry itself in terms of causation. Thus Reichenbach's *principle of the common cause*[7] states that improbable coincidences are always associated with an earlier common cause. For the purposes of the third arrow strategy, however, it is important that the order of presentation be as above. There must be a prior statistical asymmetry in the world, characterizable in noncausal terms, in terms of which causation may then be defined.

TOO FEW FORKS

I want to argue that accounts of this kind all run into the same problem: roughly speaking, there aren't enough asymmetric forks to go around. The fork asymmetry doesn't provide enough *actual* asymmetry in the world to draw the line between cause and effect where it needs to be drawn. There are two main reasons for this. The more important one is that to the extent to which there is a statistical asymmetry of this kind in the world, it is a macroscopic affair, depending on the coordinated behavior of huge numbers of microscopic components. But for the kind of reasons we have discussed in earlier chapters, the component processes themselves seem to be symmetric in the relevant respects; there seems to be no fork asymmetry in microphysics. As I'll explain, this spells defeat for the third arrow strategy.

Microphysical symmetry is not the only problem for the attempt to forge a third arrow from the fork asymmetry, however. Even in the familiar realm of macroscopic objects and processes, many common causes are simply too infrequent, too insignificant, or both, to give rise to *actual* correlations between their joint effects. We only find correlations when the causes in question are big enough and frequent enough for their effects to stand out against the background "noise," and often this simply isn't the case.

To illustrate what I have in mind here, take any common case of a cause producing a number of distinct effects—a fire producing both heat and smoke, for example. Perhaps it is true that there is a significant correlation between heat and smoke in the world. However, the fact that a given fire causes heat and smoke does not depend on this being so. Suppose the fire in question had been the only one in the history of the universe: it would still have caused both heat and smoke. More realistically, suppose that there had been a huge amount of heat and smoke produced independently by heaters of various kinds and by smoke machines—so much of each that the correlated heat and smoke produced by our fire (or even all fires) was insignificant in comparison. Intuitively, this would make no difference at all to the fact that the given fire is the cause of both heat and smoke. In other words, the causal facts seem to be quite insensitive to the amount of *actual* correlation, of the kind required for the fork asymmetry.

Construed in terms of actual correlations in the world, then, the fork asymmetry seems to provide little handle on the cause-effect asymmetry as such. There are many common causes whose effects are not significantly correlated *in practice*. So the fork asymmetry provides at best a partial distinction between cause and effect. Even if all nonlocal correlations are associated with common causes, not all common causes are associated with nonlocal correlations. There seem to be many actual cases in which an appeal to the fork asymmetry couldn't tell us whether one event is a cause or an effect of another.

Advocates of the fork asymmetry approach are likely to feel that I am being pedantic here, and misrepresenting their position. For example, it might be said that the claim is not that all common causes *actually* produce correlations between their effects, but that they all *tend* to do so—and that this is true even of kinds of causes which in practice are very rarely instantiated. The point is that if they were instantiated, they would give rise (or tend to give rise) to correlations. But these formulations introduce new modal notions—the notions of *tendency* and *what might be the case*, for example— and these notions are likely to involve temporal asymmetries of their own.

Once again, the effect is simply to shift a problem of temporal asymmetry from one place to another.

There are actually two different mistakes which advocates of the third arrow strategy are prone to make at this point. The first is what I shall call *disguised conventionalism*. The third arrow approach is supposed to be an alternative to the Humean suggestion that it is simply a matter of definition that causes occur *before* their effects. Advocates of this approach should therefore be careful not to endorse Humean conventionalism by accident, as it were, by failing to notice some subtle definitional appeal to temporal asymmetry. As we shall see, it is surprisingly difficult to avoid this mistake.

The second danger is the one mentioned in the paragraph before last. It belongs to a species familiar in philosophy in general, and endemic, as we have seen, to the study of temporal asymmetry. We might call it simply *passing the buck*. In the present context, in other words, it is the mistake of appealing to a notion whose own temporal asymmetry and orientation is no less problematic than that of causation itself. Of course, passing the buck is not always a hanging offense, in philosophy or in scientific explanation. A problem may be more tractable in a new form than in the old, for example. However, it is always important to be clear as to what an analysis achieves and what it fails to achieve. The third arrow strategy claims to explain both the intrinsic asymmetry and temporal orientation of causation in terms of an asymmetric pattern in the correlational structure of the physical world. If the project is to succeed, it is crucial that this pattern be characterized in terms which don't presuppose an asymmetry as mysterious as that of causation itself.

The next section explores these mistakes and their possible sources at greater length. It is a little more dense than usual, and readers who are confident that they themselves are not guilty of these mistakes may skip a section or two.

TWO WAYS TO MISUSE A FORK

Let's begin with a simple kind of disguised conventionalism, easily avoided. If we want to account for causal asymmetry in terms of the fork asymmetry, we might be tempted to say that causes are a certain kind of *precondition* for events, and that effects are the corresponding *postconditions,* and only then note that causes tend to be associated with correlated effects, and not vice versa. If we do that, however, we make it a conventional matter that causes precede their effects. The right approach is simply to say that causes are conditions exhibiting the required independence feature. It is then a *discovery* about the world that these are generally preconditions rather than postconditions. By beginning symmetrically, the account thus avoids conventionalism.

Most cases are not so simple, however, for they concern the nature of the fork asymmetry itself. For one thing, the fork asymmetry is normally described in terms of probabilities. Reichenbach's own account of what he calls the common cause principle runs more or less like this, for example:[8] Whenever events A and B are correlated, in the sense that $P(AB) > P(A).P(B)$, and neither A nor B is a cause of the other, then there is an earlier event C such that

1. $P(AB/C) = P(B/C).P(A/C)$

2. $P(AB/\text{not-}C) = P(B/\text{not-}C).P(A/\text{not-}C)$

3. $P(A/C) > P(A/\text{not-}C)$

4. $P(B/C) > P(B/\text{not-}C)$.

If we wish to avoid building in temporal asymmetry by fiat, however, we must be very careful about how we understand these probabilities. For example, if we thought of probability in terms of the degree of support provided by specified evidence for a given hypothesis, we would need to be careful that in describing the fork asymmetry we didn't simply incorporate a purely conventional temporal asymmetry into our conception of the relevant evidence. If we took the probability of an event at a particular time to depend on the evidence available *before* the time in question, then we should certainly suspect disguised conventionalism. Prima facie, at any rate, there would be an alternative account which simply took things the other way around.

Many contemporary philosophers reject such "evidential" accounts of probability, at least for the purposes of physics. A common view is that probability is more of an objective matter than such an approach allows. However, the objectivist accounts concerned tend to be thoroughly asymmetric in time. Objective probabilities or chances are normally taken to be "forward-looking," and dependent on the past but not the future, for example. On the face of it, this asymmetry is no less mysterious than that of causation itself. If it is an objective feature of the world, where does it come from? How is it to be reconciled with the prevailing temporal symmetry of physical theory, and more particularly with the thesis that there is no objective direction to time itself? In moving from an evidential view of probability to an objectivist one, in other words, we are likely to avoid disguised conventionalism at the cost of passing the buck.

These concerns might be avoided by using a temporally symmetric notion of probability, and a frequency account seems the best candidate. There need be no in-built temporal asymmetry in referring to correlations or patterns of association between events in the world. (Reichenbach himself thinks of the probabilities involved in the common cause principle in these terms.) Even here, however, we need to be careful with conditional probabilities. If P(A/B) were thought of in terms of the frequency with which A *succeeds* B, then again we would have an asymmetry by convention.

A concrete example may help to highlight these dangers. In an insightful paper in which he criticizes the fork asymmetry approach to causal asymmetry, Frank Arntzenius suggests as an alternative that we "relate causation to the existence of transition probabilities [in Markov processes] which are invariant under changes of *initial* distributions."[9] It seems to me that this proposal involves disguised conventionalism, however. Arntzenius is careful to allow transition probabilities in both temporal directions, and thus avoids conventionalism on that score. The problem concerns his reference to *initial* rather than *final* distributions. A simple example will provide an illustration. Suppose we have 100 identical fair coins, each of which is to be tossed once at a time t. Then the probability that an arbitrary coin will be heads after t given that it was, say, heads before t is independent of the initial distribution—that is, of the number of coins that initially show heads. Not so the reverse "transition probability," such as the probability of heads before t given heads after t: if 99 of the coins initially showed heads then this latter probability is 0.99 (assuming fair coins); if 99 showed tails then it is 0.01; and so on. Thus there is an asymmetry between forward and reverse conditional probabilities, in that only the former are independent of the initial distribution.

But what happens if we specify the final distribution instead of the initial distribution? In purely evidential terms the situation is precisely reversed. For example if we are told that after t, 99 of the coins show heads then the (evidential) probability that an arbitrary coin will be heads after t given that it was heads before t is not 0.5 but 0.99. Whereas the (evidential) probability of heads before t given heads after t is now 0.5. Thus far the direction of the probabilistic asymmetry depends on nothing more than the choice of initial rather than final boundary conditions. If Arntzenius's account is to avoid the same charge we need to be told why it cannot likewise be formulated in reverse, as it were. If it could be, the asymmetry Arntzenius claims to find would be shown to be simply conventional.

The problem is that any way of meeting this objection seems bound to pass the buck. In effect, it would be an objective justification either for preferring initial to final conditions, or for invoking a notion of probability which was sufficiently asymmetric not to embody the symmetry just noted for evidential probabilities. In either case asymmetry of causation would then rest not on Arntzenius's asymmetry in transition probabilities as such, but on whatever it was that sustained this asymmetry in the face of the above objection.

The notion of probability is not the only source of buck-passing and disguised conventionalism. Sometimes the claimed asymmetry between causes and effects is characterized in terms something like these:

(6·1)   Changing the frequency of a common cause changes the frequency of its joint effects. Changing the frequency of a common effect does not change the frequency of its joint causes.

Formulations of this kind need to be treated with particular care. This is partly because they are ambiguous. Does "changing" refer to natural variation in frequency over time, or does it refer to something like intervention, or manipulation? If the former, conventionalism and buck-passing are likely to be avoided, but the formulation offers no solution to the insufficiency of actual asymmetry in the world. What of causes whose frequency doesn't change, and is always low? Indeed, it is doubtful whether this principle captures *any* actual asymmetry. Consider two friends who maintain their friendship entirely on chance encounters. They work in the same part of town, and tend to run into each other once or twice a month. On occasions on which they do meet, their prior movements are correlated (they both happen to enter the same café at roughly the same time, for example). In months in which the total number of meetings is higher or lower than usual, the frequency of the required previous movements is also higher or lower than usual; in other words, there are more or fewer of the earlier "coincidences" which lead to one of their chance encounters. Contrary to 6·1, then, *natural* changes in the frequency of common effects are associated with changes in the frequency of their normal contributory causes.

Read in terms of intervention or manipulation, on the other hand, 6·1 is plausible enough—but it embodies the temporal perspective we have on the world as agents. As we noted earlier, agency itself seems to be temporally asymmetric: we can affect the future but not the past. Insofar as 6·1 records a genuine asymmetry, then, it is one that reflects that of agency. The best way to convince oneself of this is to note that there are conceivable agent perspectives from which the asymmetry simply doesn't hold. Consider, for example, the perspective available to God, as She ponders possible histories for the universe. For all we presently know, God may have originally had a preference for a world in which the beginning of the Third Millennium in January 2001 is marked in spectacular fashion by the occurrence of many millions of tiny fires around the globe. Among the possible histories of the world are some in which the number of individually accidental fires at that time is several orders of magnitude higher than normal. In those histories there are simply many more "accidental" conjunctions of combustibles, oxygen, and sources of ignition just prior to the given date than we would normally expect (reflecting the fact that combustibles and oxygen are among the joint causes of fires). In opting for such a history over others, God would have increased the frequency of a common effect—namely fire—and hence produced a correlation between its joint causes.

So the plausibility of 6·1 *for us* reflects the fact that our perspective as agents is very different from that of God. For us the past seems to be fixed, and inaccessible to manipulation. Insofar as 6·1 is true, moreover, its truth depends on this fact. In the next chapter I am going to argue that the fixity of the past is essentially a conventional matter—a convention we are not free to change, but a convention nevertheless. The result is that the truth of 6·1 (in its interventionist guise) is also a conventional matter, and hence a third arrow strategist who relies on it is guilty of disguised conventionalism. One might dispute the conventionality of the asymmetry of agency to avoid this conclusion, of course, but the alternative is to treat the asymmetry of agency as an objective matter, and this will amount to passing the buck.

This is by no means a comprehensive survey of the ways in which it is possible for the third arrow strategy to go off the rails in trying to make use of the fork asymmetry, and to fall into buck-passing or disguised conventionalism. To close, however, it may be useful to have an illustration of a related account of causal asymmetry which is conspicuous for the care its author takes to avoid this kind of mistake.

A FOURTH ARROW?

David Lewis, a leading contemporary Princeton philosopher, proposes to analyze causation in terms of what philosophers call "counterfactual conditionals," or simply "counterfactuals." These are claims about what would have happened in circumstances which are known not to have happened, or at least not known to have happened; typically, they take a form such as "If it were (or had been) the case that P, then it would be (or would have been) the case that Q." Roughly speaking, Lewis's suggestion is that to say that an event A caused an event B is to say that if A had not happened, B would not have happened.

Lewis points out that such an account may trace both the asymmetry and the predominant temporal orientation of causation to an asymmetry concerning counterfactuals. He describes this asymmetry as follows:

> The way the future is depends counterfactually on the way the present is. If
> the present were different, the future would be different. ... In general the

way things are later depends on the way things are earlier. Not so in reverse. Seldom, if ever, can we find a clearly true counterfactual about how the past would be different if the present were somehow different.[10]

By way of illustration, consider the claim that if my parents had not met in the 1940s, this book would not have been written. This is highly plausible, but think about what happens if we try to make the dependence go the other way: "If *Time's Arrow and Archimedes' Point* had not been written, Huw Price's parents would not have met in the 1940s." The oddity of this claim is a very general feature of counterfactuals which run from future to past in this way.

As Lewis emphasizes, however, his suggestion doesn't immediately explain causal asymmetry. It simply shifts the problem from one place to another. To explain the causal asymmetry we now need to explain the asymmetry of counterfactual dependence. Lewis tries to do this in terms of what he calls the asymmetry of overdetermination—a feature of the world closely related to the fork asymmetry. In effect, then, Lewis interposes a fourth arrow, an asymmetry of counterfactual dependence, between the arrow of causation and the arrow constituted by the fork asymmetry. He himself is not seeking to put the weight of the account of causal asymmetry on this fourth arrow. Its role is simply to transfer the burden from one place to another. The mistake he thus avoids is that of *merely* passing the buck, that of simply moving the problem from one place to another.

We shall return to the asymmetry of counterfactual dependence in the next chapter. I want to try to show that explained in a very different way from Lewis's approach, it provides a particularly clear elucidation of the subjective ingredient that I take to be an essential part of an adequate account of the asymmetries of causation and dependence. For the moment, however, the lesson to be learned from Lewis's careful approach is that modal "arrows" need to be treated with a great deal of caution. Their own asymmetry is likely to be as problematic as that of causation itself. In particular, third arrow strategists should not rest content with modal readings of the fork asymmetry. Insofar as the world exhibits an asymmetry of any such kind, that fact will be as much in need of explanation as the causal arrow itself.

THE SYMMETRY OF MICRO-FORKS

We now turn to the major difficulty which confronts the attempt to ground causal asymmetry on the fork asymmetry. Here too Lewis's account provides a useful illustration, but this time of vice rather than virtue. Lewis's asymmetry of overdetermination consists in the fact that events typically have very few earlier determinants, but very many later determinants. (In Lewis's terminology, an event A is a determinant of an event B if A is just sufficient to ensure that B, given the laws of nature.) As Lewis puts it:

> Whatever goes on leaves widespread and varied traces at future times. Most of these traces are so minute or so dispersed or so complicated that no human detective could ever read them; but no matter, so long as they exist. It is plausible that very many simultaneous disjoint combinations of traces of any present fact are determinants thereof; there is no lawful way for the combination to have come about in the absence of the fact. (Even if a trace could somehow have been faked, traces of the absence of the requisite means of fakery may be included with the trace itself to form a set jointly sufficient for the fact in question.) If so, the abundance of future traces makes for a like abundance of future determinants. We may reasonably expect overdetermination toward the past on an altogether different scale from the occasional case of mild overdetermination toward the future.[11]

Lewis himself refers to the apparent asymmetry of radiation as a special case of the asymmetry of overdetermination:

> There are processes in which a spherical wave expands outward from a point source to infinity. The opposite processes, in which a spherical wave contracts inward from infinity and is absorbed, would obey the laws of nature equally well. But they never occur. A process of either sort exhibits extreme overdetermination in one direction. Countless tiny samples of the wave each determine what happens at the space-time point where the wave is emitted or absorbed. The processes that occur [in our world] are the ones in which this extreme overdetermination goes toward the past, not those in which it goes toward the future. I suggest that the same is true more generally.[12]

However, recall what we learned about radiation in chapter 3. The apparent temporal asymmetry of radiation is somewhat misleading. It stems not from the processes of radiation themselves, but from the fact that there is an imbalance between sources and sinks. The universe as we know it contains big coherent sources of radiation, but no corresponding big sinks. All the same, when it comes to representing radiation "atomistically," that is, as the sum of its many individual components, the asymmetry disappears. It makes no difference at all whether we think of the transmitters as tiny sources of outgoing waves, or the absorbers as tiny sinks of incoming waves. The two mathematical representations are equivalent.

To illustrate the effect of this point, imagine a tiny source of radiation, a single atomistic source (call it *i*). Using the source-based representation of the radiative field concerned, we can in principle represent the difference between the presence and absence of i at a given time t. Because i is a source rather than a sink, the difference would show up after t, rather than before. At times after t, in other words, we could in principle determine whether i had been present at t by the presence or absence of an outgoing ring of correlated changes in the field. In order to "see" this pattern of correlations, however, we would need to be able to "correct for the background"—in other words, to know the state of the field immediately before t (its state insofar as it does not depend on i, in effect). But if we know this, and know that i is the only variable, then for us i is overdetermined by the later state of the field. It is just like looking at a segment of the outgoing ripple on the pond, and thence being able to tell that a disturbance occurred at a certain time at a certain place on the surface. (The difference is that in the latter case we seem to get by without knowing the background conditions. More on this below.)

However, what applies with respect to the atom-source i applies equally with respect to an atom-sink i^*. In other words, using the sink-based representation of the radiative field concerned, we could in principle represent the difference between the presence and absence of i^* at a given time t^*. Because i^* is a sink rather than a source, the difference would show up before t^*, rather than after. At times *before* t^*, in other words, we might in principle determine whether i^* is present at t^* by the presence or absence of an incoming ring of correlated changes in the field. In order to "see" this pattern of correlations, however, we would need to be able to "correct for the background"—in other words, to know the state of the field immediately *after* t^* (its state insofar as it does not depend on i^*, in effect). If we know this, and know that i^* is the only variable, then for us i^* is overdetermined by the earlier state of the field.

Hence in the case of radiation there is no asymmetry of overdetermination at the micro level. The familiar macroscopic asymmetry arises because of the imbalance between big coherent sources (of which there are many) and big coherent sinks (of which there are none, so far as we know). Note that the real importance of size is that it enables one to ignore the background. Familiar coherent sources are so statistically improbable that the background becomes insignificant. This is how we are able to determine what happened at the center of the pond without explicit knowledge of the initial conditions. As long as the initial conditions were not themselves the product of such a coherent source, their contribution is effectively irrelevant. In this way we get Lewis's kind of overdetermination, viz., the ability to infer the occurrence of a core event simply from a knowledge of any one of its many traces.

One way to appreciate the problem for Lewis's account is to imagine the power of a wave transmitter progressively reduced. As the signal fades, it becomes more and more difficult to distinguish from the background noise. Two kinds of extra information can help at this point. One is the kind of information we get by sampling the field over a wider region of space. The other is the kind of information we get by being told more about the state of the field immediately prior to transmission. In the limit we need complete information of one kind or another, or some appropriate mix, just as in the atom-source case. Overdetermination in Lewis's sense is thus a matter of degree, and "fades out" as we approach the micro level. However, it seems quite implausible to suggest that the same is true of the asymmetries of dependence and causation.[13]

These points seem to apply with equal force to other attempts to ground the asymmetry of causation on the fork asymmetry. There seems to be no fork asymmetry at the microscopic level. This point has been obscured by a number of factors, I think. One important influence has been the intuitive appeal of μInnocence. As we saw in the previous chapter, we are very strongly attached to the idea that an interaction in the past and an interaction in the future have very different significance: only in the first case do we expect the systems concerned to be correlated. However, we saw that once μInnocence is properly distinguished from the macroscopic asymmetry associated with the presence of entropy-reducing correlations toward the past but not toward the future, μInnocence itself seems to have no observable physical basis. Far from leading us to a physical asymmetry which could play the role of the third arrow, then, μInnocence merely compounds the mystery of the asymmetries of causation and dependence. It is a powerful intuitive asymmetry which seems to have no grounding in the physical world.

The lack of a microscopic fork asymmetry has perhaps also been obscured by a tempting fallacy. Suppose we consider microevents P, F_1, and F_2, as in Figure 6.2. It will typically be the case that if P is held fixed then F_1 and F_2 may be correlated with one another; if F_1 had been different, F_2 would have been different. Given conservation of momentum, for example, holding P fixed may ensure that the momentum of F_1 is related to that of F_2. But this is not the fork asymmetry at work, as we see if we note that precisely the same obtains in reverse. If we hold F fixed, then the same will apply to P_1 and P_2. Temporal direction is irrelevant.

Thus the fork asymmetry we actually find in the world is a matter of what might be called gross forks: it disappears if we focus closely on the microstructure of the physical processes in which it shows up. By way of comparison, think of what happens when you look very closely at a computer screen image: the pictorial content of the image disappears, in favor of a

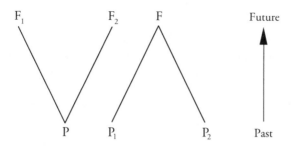

Figure 6.2. A symmetric arrangement of micro-forks.

pattern of individual pixels. The image is not pictorial "all the way down," as we might say. The same seems to be true of the fork asymmetry. It depends on the ordered alignment of vast numbers of microscopic events, and hence is simply not the sort of feature of the world which could be manifest when the numbers involved are too small.

At the microscopic level, then, there seems to be no fork asymmetry. Does this mean that at the micro level we have fork correlations in *both* temporal directions? Well, yes and no. As we saw above, such micro-forks show up in either temporal direction, but only if the background is thought of as perfectly fixed. Thus they differ from macro-forks, whose signal strength is so far above the background that the background may be ignored. Micro-forks are what we have in Figure 6.2, in effect.

Why does this matter, for someone who wants the fork asymmetry to provide the missing third arrow that underlies the asymmetry of causation? Simply because it seems to imply that when we look in detail at the individual microscopic processes of which the world is composed, *there is no asymmetry of cause and effect.* It means that at least at this level, there is no right answer to the question as to which of a pair of events is the cause and which the effect. Did the incoming photon cause the electron to leave the electrode, for example, or was it the other way around? If causal asymmetry requires the fork asymmetry, and there is no fork asymmetry at the microscopic level, then neither answer seems appropriate.

Of course, some philosophers and many physicists will say that they see nothing wrong with this conclusion—that it simply confirms what they already suspected, namely that there is no such thing as real causation in the physical world (or at least the microphysical world). I shall come back to this response in a moment. For the moment, the important thing is that for those philosophers and physicists who do take causation—ordinary, asymmetric causation—to be a real ingredient of the physical world, the third arrow strategy is in trouble. Initially, the fork asymmetry seemed to provide the only plausible candidate for an objective third arrow. In practice, however, it turns out to be woefully inadequate to the task at hand. It disappears where it is needed most, in a sense, at the level of the microscopic substructure we take to underlie the familiar world of ordinary experience.[14]

TWO EXTREME PROPOSALS

We have been exploring the suggestion that the temporal asymmetry of causation might be explicable in terms of a de facto physical temporal asymmetry. However, we have concluded that this strategy is unable to make sense of our intuitions concerning dependence in microphysics. Hence we are confronted once again with a puzzle that has troubled physics at least since the late nineteenth century. On the one hand, the micro world seems predominantly symmetric in time, in the sense that individual processes that occur with one temporal orientation also occur with the opposite orientation. On the other hand, we want to say that microphysical events have causes and effects, and depend on earlier states of affairs in a way in which they do not depend on later states of affairs. These claims are difficult to reconcile, however. One pleads for temporal symmetry at the fundamental level of reality, while the other insists that there is a very basic asymmetry.

There are two extreme responses to this puzzle. One option is simply to banish the notion of causality from physics, or at least microphysics. This is a course once famously advocated by Bertrand Russell:

> All philosophers, of every school, imagine that causation is one of the fundamental axioms or postulates of science, yet, oddly enough, in the advanced sciences such as gravitational astronomy, the word "cause" never occurs. ... It seems to me ... that the reason physics has ceased to look for causes is that, in fact, there are no such things. The law of causality, I believe, like much that passes muster among philosophers, is a relic of a bygone age, surviving, like the monarchy, only because it is erroneously supposed to do no harm.[15]

Russell's view would today be called an "eliminativist" or "error theory" view of causation. It is the view that folk or philosophical theories of causation have simply been shown to be *false* by the march of science. As I noted earlier, views of this kind are quite popular among physicists (even if they don't always practice what they preach, continuing to be influenced by the asymmetric aspects of the folk intuition, for example).

What's wrong with such a view? It is counterintuitive, of course, but we knew this already. After all, it's an appeal to the thought that folk intuitions are sometimes shown to be mistaken by physics. All the same, it is important to be clear which intuitions are actually at stake. The view would be less unattractive if it simply challenged our intuitions about causality in microphysics. However, the damage seems unlikely to be contained to the micro world. We like to think that at least in physics, big things are collections of little things—and in particular, that macroscopic causation is made up of a lot of microscopic causation. One aspect of this intuition is that causal connections between temporally separated events decompose "horizontally" into chains of more immediate causal connections; another is that these immediate connections, if not already primitive, decompose "vertically" into a complex of microphysical causal relations. So if there is no temporal asymmetry in these microphysical relations, it is hard to see how putting a lot of them together could make any difference.

Thus if we reject causation in the micro world, there are good prospects of an argument to the conclusion that all talk of causation is ultimately groundless. With causation will go asymmetric dependence, and apparently our right to the view that our present actions affect the future but not the past. This is not an option to be taken lightly, in other words. We may eventually conclude that physics requires us to jettison our customary talk of causation and dependence, but we should be sure that we have exhausted less radical possibilities.

The conventionalist strategy provides one such possibility. It proposes to explain our ordinary causal talk in such a way that it does not conflict with the temporal symmetry of physics. It argues that the asymmetry of dependence and causation is a conventional matter: built into these notions by definition, so to speak. If this strategy works, then eliminativism will turn out to have been a reaction to a nonexistent problem (at least in so far as it is motivated by the problem of causal asymmetry). In this context, it is interesting to note that even Russell soon finds himself explaining the folk use of the notions of cause and effect in conventional terms. At one point he considers the objection that "our wishes can *cause* the future, sometimes, to be different from what it would be if they did not exist, and they can have no such effect on the past." He replies that "this … is a mere tautology. An effect being *defined* as something subsequent to its cause, obviously we can have no *effect* upon the past."[16]

Earlier I described the Humean version of the conventionalist strategy. In the next chapter I want to propose a rather more sophisticated version. Its effect, I think, is to make the eliminativist's extreme solution quite unnecessary. A well-motivated conventionalist view shows that physics need not concern itself especially with cause and effect, that these notions are not part of the proper subject matter of physics. This is very different from the view that physics shows that ordinary talk of cause and effect is *in error.*

Before we turn to the defense of an improved conventional solution, let me mention a second extreme response to the apparent conflict between the symmetry of physics and the asymmetry of causation, which is to deny the temporal symmetry of the micro world. According to this view, we find symmetry in the micro world only by ignoring a real and fundamental constituent of the world, namely the causal relations between its events. Physics itself may be symmetric at the microscopic level, but there is another aspect or level of reality which is asymmetric—the level of dependence, causality, and the like.

We might call this the "hyperrealist" view of causation. It takes causation to be something over and above the concerns of physics—something as real as the aspects of the world with which physics is immediately concerned, but not reducible to those aspects. According to an account of this kind it is not surprising that we do not find causal asymmetry within microphysics. Strictly speaking, we don't find it within any sort of physics.[17]

The main difficulty with this view is that in putting causation and dependence beyond physics, it threatens to make them inaccessible and irrelevant. The inaccessibility in question is epistemological: it seems to be a consequence of this view that we will simply have no way of *knowing* whether our ordinary ascriptions of the terms cause and effect are correct or back to front. Perhaps the past actually depends on the future, and not vice versa— but how could we tell? The argument in the previous section suggests that we will not be able to settle the matter by looking at microphysics. The only temporal asymmetry that could possibly constitute evidence one way or the other is the macroscopic thermodynamic asymmetry, and here, as we have seen, the best explanation seems to lie not in causal connections between particular events (a path that would in any case lead us back to microphysics) but in cosmological constraints on the boundary conditions of the universe. Hence it seems unlikely that the thermodynamic asymmetry could provide evidence concerning such extraphysical ingredients of the world.

The difficulty of knowing anything about causal relations of this kind is only half the problem, however. An advocate of the hyperrealist view of causation needs to explain how the existence of such features of the world connects with our capacities to affect the world. And yet in being distinct from the facts of physics, such extraphysical causation seems remote from everything of any significance to our life as agents and actors in the world. After all, what difference would it make to us if there were no extraphysical states of affairs? Wouldn't everything seem just as before?[18]

In sum, we looked for causal asymmetry in the physical world, but couldn't find it. The eliminativist recommended that we bite the bullet, and accept that all our talk of asymmetric causation is mistaken. I advised caution, however. Until we have a better idea of what we actually do with causal talk—how it works, what it does for us, whether we should *expect* to find it mirrored in physics—throwing it overboard seems a rather hasty solution to a poorly defined problem. The causal hyperrealist agrees with me on this point, but rushes equally hastily in the opposite direction, embracing an element of reality over and above what physics describes. Here I advised skepticism as to whether such an approach can actually make sense of the causal talk and intuitions it seeks to save from the eliminativist's clutches.

THE PERSPECTIVAL VIEW

The approach I recommend is much more modest and pragmatic. Perhaps causal asymmetry isn't really in the world at all, but the *appearance that it is* is a product of our own standpoint. Perhaps it is like the warmth that we see when we look at the world through rose-tinted spectacles. Or, to use the analogy that Kant used for very much this purpose, perhaps it is like the apparent motion of the heavenly bodies, which the Copernican Revolution shows us to be a product of the motion of the viewpoint on which we stand. An obvious advantage of this approach is that it promises to make do with very much less *actual* asymmetry than the third arrow strategy requires. Just as the *apparent* motion of the stars is very much more widespread than the *real* motion in terms of which Copernicus accounts for the appearances, so the *apparent* causal asymmetry might be expected to be very much more widespread than the actual asymmetry which explains it. This is a very economical approach, in other words, well placed to handle the shortages of actual asymmetry which we saw to plague the third arrow strategy.

More on this theme in a moment. First, however, an obvious question is how this approach differs from eliminativism—from the view that science simply shows causal talk to be *mistaken?* The difference lies mainly in the approach's attitude to the discovery that causal talk is perspectival. The eliminativist simply assumes that causal talk is trying to do what physics does—"to describe the world," perhaps—and therefore rejects causal talk when it is discovered that there is nothing of the appropriate sort in the world. The present view is more subtle, and reads the discovery of the perspectival character of causal talk as the discovery of a difference between causal talk and physics—the discovery that they are in different lines of work, in effect. The eliminativist's mistake is to judge causal talk by the wrong standard, a standard which the perspectival account shows to be inapplicable.

This may sound like some sort of semantic sleight of hand, but it rests on a simple and plausible point, with some familiar applications. In chapter 1, for example, I described the old philosophical controversy about the status of the distinctions between past, present, and future. One side takes the view that these distinctions represent objective features of reality, the other that they rest on a subjective feature of the perspective from which we view the world. On the latter view, the notion of the present, or *now*, is perspectival in much the way that *here* is. Just as there is no absolute, perspective-independent *here* in the world, so there is no absolute, perspective-independent *now*. Obviously a view of this kind does not have to say that our ordinary use of "here" and "now" is mistaken, however, and in need of the eliminativist's radical surgery. Once we understand the perspectival character of these terms, we see that their ordinary use does not involve a mistaken view of the nature of reality: it doesn't presuppose that there is an objective *here* and an objective *now*. (The mistake, if there is one, lies in a philosophical misinterpretation of the ordinary practice.)

In general, then, the viewpoint I am recommending is very different from eliminativism. What makes the perspectival approach particularly appealing in the present case is that we ourselves are strikingly asymmetric in time. We remember the past and act for the future, to mention but two of the more obvious aspects of this asymmetry. It does not seem unreasonable to expect that the effects of this asymmetry might have become deeply entrenched in the ways in which we describe our surroundings; nor that this entrenchment should not wear its origins on its sleeve, so that it would be easy to disentangle its contribution from that part of our descriptions we might then think of as perspective-independent. After all, there is a familiar model for this in the case of the familiar secondary qualities—colors, smells, tastes, and so on. In these cases the familiar categories are now recognized to be in part a product of the peculiarities of our own sensory apparatus. However, the appreciation of this point was far from a trivial intellectual achievement, as we saw in chapter 1; early in the seventeenth century it was a controversial matter, fit to be argued by a thinker of Galileo's stature. And the precise nature and extent of the secondary qualities remain very much matters for contemporary philosophical debate.[19]

The analogy with the familiar secondary qualities thus serves to meet two objections to the proposal that causal asymmetry might be perspective-dependent. The first objection is that if causal asymmetry were perspectival, this would already be obvious to us, and hence we would be less inclined to regard the cause-effect distinction as an objective matter. The second objection follows on, claiming that this view would therefore conflict with the practice of physicists, who do treat this distinction as having objective significance.

The case of the ordinary secondary qualities has two lessons: First, that it is far from easy to distinguish appearance from reality, to decide how much of the manifest image of the world is a product of our own perspective; and second, even when the influences of our perspective are discerned, it is far from clear what we should do about them—far from clear, in particular, that the right thing is to abandon the language of the secondary qualities, at least in scientific discourse. It is not clear that this is a serious option.[20]

What feature of our perspective could it be that manifests itself in the cause-effect distinction? The most plausible answer is that we acquire the notion of causation in virtue of our experience as agents. Roughly, to think of A as a cause of B is to think of A as a potential means for achieving or bringing about B (or at least making B more likely). One of the earliest statements of a view of this kind is in the posthumously published work of the brilliant young Cambridge philosopher Frank Ramsey (1903–1930). In the later and less well known of his two papers on laws of nature, Ramsey makes a number of remarks about causation and its direction. He links our notions of cause and effect to the agent's point of view, saying that "from the situation when we are deliberating seems to ... arise the general difference of cause and effect."[21]

This general approach has become known as the *agency,* or *manipulability,* theory of causation. It has not been particularly popular among philosophers, although the standard objections to it seem to be rather overrated. They turn out to be closely analogous to, and no more forceful than, a range of objections that might be made against quite standard philosophical theories of color.[22] However, the point I want to emphasize here is not that the agency account of causation lacks the disadvantages with which it has usually been saddled, but rather that such an account is exceptionally well placed to explain the nature of causal asymmetry, and its prevailing orientation in time. It is able to say that the asymmetry of causation simply reflects that of the means-end relation. Causes are potential means, on this view, and effects their potential ends. The origins of causal asymmetry thus lie in our experience of doing one thing to achieve another—in the fact that in the circumstances in which this is possible, we cannot reverse the order of things, bringing about the second state of affairs in order to achieve the first. This gives us the causal arrow, the distinction between cause and effect. The alignment of this arrow with the temporal arrow then follows from the fact that it is normally impossible to achieve an *earlier* end by bringing about a *later* means.

This is just the beginning, of course. It needs to be explained how the means-end relation comes to have these characteristics—in particular, how it comes to have such a striking temporal orientation. A natural suspicion is that such an explanation would itself need to appeal to the asymmetry of cause and effect, thus invalidating the proposed account of causal asymmetry. I shall consider this objection in a moment, but first let me emphasize again the basic motivation and attractions of this perspectival approach. Remember that the problem was that we couldn't find enough temporal asymmetry "out in the world" to form the basis for our intuitive judgments about what causes what. The perspectival solution to this problem is to say that the asymmetry of causation is a kind of projection of some internal asymmetry in us, rather than a real asymmetry in the world. In effect, the reason that we see asymmetry everywhere we look is that we are always looking through an asymmetric lens.

In order to develop this suggestion, we need to find some asymmetric feature of our own circumstances to play the role of the asymmetric lens—and this feature has to be connected with our use of notions such as causation and dependence in the right kind of way. I have suggested that the crucial thing is that we are *agents:* we deliberate about our actions, and this process itself is both intrinsically asymmetric and strongly oriented in time. (The asymmetry of agency was one of the three temporal asymmetries we distinguished at the beginning of this chapter.) With this foundation, the perspectival view becomes the claim that our talk of causation and dependence is a projection from the kind of perspective we have as agents in the world.

A common concern about this approach is that it involves some sort of vicious circle. In particular, opponents charge that the asymmetry of agency depends on that of causation: the reason we can only act for later ends is that causes always precede their effects. In the next section I outline a response to this challenge. This section is a little more dense than usual, however, and readers should feel free to skip to the end of the chapter.

In the next chapter I want to try to approach the perspectival view by a more thorough and more perspicuous route. As we shall see, it turns out to be helpful to focus more on the asymmetry of dependence than on causation itself. In effect, however, I shall be presenting the agency account of causal asymmetry as a more sophisticated version of Hume's conventionalist strategy. I shall argue that it is our de facto temporal orientation as agents that requires that we choose the relevant convention as we do. As we shall see, this provides a kind of surrogate objectivity, which does much to blunt the objections to conventionalism we mentioned earlier in the chapter— in particular, the argument that conventionalism cannot make sense of the sheer objectivity of the fact that the future depends on the past, but not vice versa. Most important, however, the task of the next chapter will be to clarify the status of some of the intuitions with which we began; to decide, for example, whether the intuition that the state of our incoming photon is independent of the orientation of our sunglasses is *merely* a consequence of the subjectivity of the asymmetry of dependence, or reflects some deeper and more problematic prejudice in contemporary physics.

ESCAPING A CIRCLE, PROJECTING AN ARROW

If the perspectival approach is to appeal to the fact that we are agents, it needs to explain the asymmetry of agency: in other words, how the means-end relation comes to have such a striking temporal orientation. As we saw, a natural suspicion is that this explanation will need to appeal to the asymmetry of cause and effect, thus rendering circular the proposed account of causal asymmetry. However, it seems to me that this suspicion is bound to be unfounded, if the alternative is that causal asymmetry is to be explicated in terms of some objective asymmetry in the world, as it is by the third arrow strategists. For in this case the latter asymmetry will itself be available to explain the asymmetry of the means-end relation. Reference to causation in such an explanation will simply be construed as indirect reference to the underlying objective asymmetry, so that the explanation does not make any essential or ineliminable reference to causation at all.

In other words, here is a plausible strategy for explaining the asymmetry and temporal orientation of agency. First, explain it making full use of the ordinary notions of cause and effect. Then rewrite the explanation, so that at those points at which it appeals to the asymmetry of cause and effect, it refers instead to the kind of objective physical asymmetry that third arrow strategists take to constitute causal asymmetry. The result will be an explanation of the relevant features of agency which does not itself appeal to causal asymmetry.

It is a familiar point that we know more about the past than we do about the future, and that we normally deliberate and act for the sake of the future but not the past. Moreover, it seems natural to explain these asymmetries in terms of that of causation: essentially, in terms of the observation that knowledge is an effect and actions are causes. Developing these ideas may be expected to yield a formal model of what is essential to our status as knowers, deliberators, and agents in the world. According to the agency theory, it will be this model that a creature must instantiate if it is to develop and possess the concept of causation—for causal concepts depend on these features in much the same way that color concepts depend on our color vision.

As it stands, the model will make free use of the notions of cause and effect. However, it seems that we will be free to regard references to these notions as placeholders for references to the physical asymmetries in the world on which the existence of such asymmetric entities actually depends. Indeed, a good indication of how this stage of the account might go is to be found in the approach to various temporal asymmetries recommended by Paul Horwich. Horwich suggests that the fork asymmetry underpins what he calls the *knowledge asymmetry:* the fact "that we know more about the past than we do about the future." The connection turns on the fact that "the processes that give us knowledge about the past are typically [future-directed open forks]."[23] Notice that there is no talk of causation here; what we use instead are the notions of process and correlation. Hence Horwich may legitimately go on to argue that the knowledge asymmetry underlies asymmetries of explanation and hence of causation itself. At this stage the account he offers differs from the agency theory of causation in several respects, but these differences need not concern us here. The important point is simply that Horwich's project illustrates the way in which the existence of knowledge, deliberation, and action may plausibly be held to rest on physical asymmetries in the world, and hence not to depend in any essential way on concepts such as causation (which it is suggested they be invoked to explain).

Why will such an approach not be vulnerable to a version of the objection I have raised to the fork asymmetry-based approaches, namely that they cannot account for the asymmetry of microphysical causation? Because agents are essentially macroscopic, and depend on the very thermodynamic asymmetry which is the source of the various physical asymmetries to which third arrow strategists appeal. In other words, this route to an explanation of the asymmetry and temporal orientation of agency would invoke only the "correct" part of its opponents' theory of causal asymmetry. As I noted earlier, accounts of this kind are very economical. A small objective change in one's viewpoint can alter the apparent character of one's entire surroundings. The agency theory of causation thus offers the basis for an account of causal asymmetry which promises to succeed not only where the third arrow strategy tends to do well, in common and macroscopic cases, but also where it fails, in rare and microscopic cases.

The great disadvantage of the approach may seem to be that it makes causal asymmetry an anthropocentric matter. My view is that we should acknowledge this consequence, but deny that it is a disadvantage. Its effect is merely to put causation in its proper metaphysical perspective, as something like a secondary quality.[24] As in the case of the more familiar secondary qualities, the shift in perspective may make us feel metaphysically impoverished, in losing what we took to be an objective feature of the world. The feeling should be short-lived, however. After all, if what we appear to have lost was illusory anyway then our true ontological circumstances are unchanged—and yet we will have made a direct gain on the side of epistemology, as we came to understand the source of the illusion.

SUMMARY

1. We have very powerful intuitions concerning the asymmetries of causation and dependence, even about very simple physical cases.

2. These intuitions cannot be dismissed as being "merely subjective," and hence of no relevance to physics, since they seem to influence current theory in very central ways.

3. In the attempt to find an objective physical basis for these intuitions, however, the fork asymmetry seems to provide the only plausible candidate, and it turns out to be insufficiently general. The main problem (though not the only one) concerns its systematic failure in microphysics. The model of radiation and thermodynamics suggests that the fork asymmetry is essentially macroscopic.

4. This conclusion has been obscured by the appeal of μInnocence, and also by the fact that some attempts to invoke the fork asymmetry actually rely either on disguised conventionalism or on some form of buck-passing. We need to be particularly suspicious of appeals to further modal notions, such as probability and counterfactual dependence. Their own temporal asymmetry is likely to be as mysterious as that of causation itself.

5. Eliminativism and hyperrealism about causal asymmetry are (somewhat heroic) options at this point. The first takes ordinary causal talk to be shown to be massively mistaken by physics, while the second takes causal talk to be referring to a reality over and above that of physics.

6. According to the perspectival view, however, these drastic measures are unnecessary. It is only if we fail to see the perspectival character of causal talk that we think its asymmetries conflict with physics. On the perspectival view, causal asymmetry reflects an asymmetry in us, not an asymmetry in the external world. This avoids the problems of approaches based on the fork asymmetry. In accounting for appearances, a little asymmetry goes a very long way.

7. The agency theory of causation offers the most promising basis for such a perspectival view. The process of deliberation seems to have the right kind of intrinsic asymmetry, and the right kind of temporal orientation. Perhaps most important of all, the agent's perspective is one we all share, all of the time; small wonder then that it colors our view of the world, in ways we find hard to detect.

7

←——————→

Convention Objectified and
the Past Unlocked

CHAPTER 6 closed with the suggestion that causal asymmetry might be conventional, or perspectival—not an objective aspect of the world, but a kind of projection of our own internal temporal asymmetry. As I noted, this diagnosis is likely to appeal to many physicists, who are often inclined to dismiss causation as a subjective matter. However, I suggested that physicists who take this view don't seem to practice what they preach, in that they continue to allow their theoretical work to be guided by the familiar asymmetric intuitions about causation and dependence. One goal of this chapter is to try to clarify this issue, and to show that despite the subjectivity of our ordinary causal notions, there really are objective matters at stake here. There is an objective possibility, strongly favored on symmetry grounds, which physics has largely overlooked.

Before we come to those issues, however, I have to try to satisfy some opponents on the opposite flank. Philosophers are much more inclined than physicists to treat causation as a fully objective matter. Accordingly, many philosophers will feel that a conventionalist approach cannot do justice to the apparent objectivity of causal asymmetry. The first task of the chapter is to respond to opponents of this kind.

For both purposes it turns out to be very helpful to focus not on the temporal asymmetry of causation itself, but on that of dependence—in other words, to think not so much about why causes occur before their effects, as about why the future depends on the past (and not the other way round). Moreover, it turns out to be useful to interpret the latter notion in terms of counterfactual conditionals: in terms of the idea that if the past had been different, the future would have been different, but not vice versa. As we shall see, this allows us to characterize both the relevant asymmetry and its conventional basis in a particularly sharp form, and also to keep a useful distance between the convention itself and those features of our own situation on which it rests. In a sense, then, the project is to transcribe the subjective viewpoint outlined at the end of the previous chapter into a new key, and thereby to exhibit its strengths to opponents from both the above camps.

ASYMMETRY CONVENTIONALIZED

Let's return to our photon, passing through sunglasses after a long journey from a distant star. The intuition I appealed to in chapter 6 was that the state of the incoming photon does not depend on the fact that it is going to pass through a polarizing lens in the near future, but may depend on the fact that it passed through such a lens in the distant past. This intuition is temporally asymmetric, and puzzling, as usual, because the laws of physics seem to be symmetric in all relevant respects.

We might conclude that the intuition concerned is simply mistaken, but we have seen that this would be too hasty. Perspectivalism offers a less drastic way to resolve the puzzle, suggesting that the asymmetry stems from us, and hence that the fact that it has no basis in the external world is not a cause for concern. In effect, this is the Kantian move I described in chapter 6. We explain the apparent objectivity of the motion of the heavens in terms of a real motion of our standpoint.

This is the idea we want to explore, and a useful approach is to do away with explicit talk of dependence, by trying to couch the intuitive asymmetry in counterfactual terms. Intuitively, it seems true that

(7·1)   If the *distant* polarizer had been oriented differently, the state of the incoming photon might have been different (or the photon might not have reached this region at all),

and yet false that

(7·2)   If the *local* polarizer had been oriented differently, the state of the incoming photon might have been different (or the photon might not have reached this region at all).

This counterfactual formulation provides a useful representation of at least one aspect of the intuitive asymmetry in the case in question, and many similar cases. As it stands, however, it leaves the asymmetry no less mysterious. It is still puzzling as to what in the world could sustain such a difference between the past and the future. (In a sense things have become more puzzling, because of the way in which the counterfactuals appear to direct us to nonactual states of affairs.)

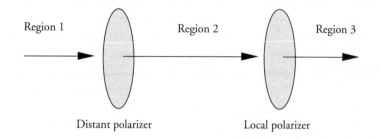

Figure 7.1. The path of the photon.

The possibility we want to explore is that the asymmetry concerned doesn't really have anything to do with the photons themselves, but is simply buried in the content of counterfactual claims. One story about the meaning of counterfactual claims goes something like this. When we assess a counterfactual of the form

(7·3) If X had happened at time t_1, Y would have happened at time t_2

we consider the consequences of a hypothetical alteration in the course of events. We "hold fixed" the course of events *prior* to t_1, assume that X occurs at that time, and consider the course of events that *follows*. Roughly speaking, we take the conditional to be true if Y at t_2 is a consequence of the joint assumption of the actual history prior to t_1 and the occurrence of X at t_1.[1]

This isn't the only suggestion philosophers have made about the meaning of counterfactual claims, but it is the most popular approach, and it does demystify the temporal asymmetry involved in the contrast between 7·1 and 7·2. To make claim 7·1 is to say that if we hold fixed the history of the photon *before* it reaches the distant sunglasses, but assume the sunglasses to be differently oriented, we may derive different consequences for the state of the photon just before it reaches the Earth. Whereas to make claim 7·2 is to say that if we hold fixed the history of the photon *before* it reaches the local sunglasses, but assume them to be differently oriented, we may derive different consequences for the state of the photon before it reaches the Earth. The latter claim is clearly false, but for a rather trivial reason: the state of the photon doesn't change under the assumed circumstances, because the assumption includes the supposition that it doesn't!

More important, it is now easy to see that the case involves no real violation of temporal symmetry. Consider Figure 7.1: The counterfactual 7·1 involves the suppositions that Region 1 is held fixed, and that the setting of the distant polarizer is varied. Under these conditions it is claimed to follow that the state of the photon in Region 2 varies. The temporal image of this supposition would be that Region 3 is held fixed, and the local polarizer setting varied. However, this is not at all what is involved in proposition 7·2 (the claim that the state of the photon before it reaches the Earth does not depend on the local setting). Rather, the latter claim involves the supposition that Region 2 (and presumably 1) are held fixed, and the local polarizer allowed to vary. Under these conditions it is of course a trivial consequence that Region 2 doesn't change.

In other words, this account of the significance of counterfactual claims seems to dissolve the tension between the asymmetry of dependence—at least as revealed in the contrast between 7·1 and 7·2—and the apparent symmetry of the physical phenomena themselves. It does this by making the asymmetry simply a conventional matter. Why doesn't the present affect the past? Because to say that the present affected the past would just be to say that if the present had been different the past would have been different. And that in turn is just to say that if we suppose the present to be different, while the past is the same, then it will follow that the past is different. This is untrue, of course, but simply on logical grounds. No physical asymmetry is required to explain it.

But is this the right way of dissolving the tension? An obvious objection is that it makes our ability to affect the future but not the past a terminological matter. After all, couldn't we have adopted the opposite convention, assessing counterfactuals by holding the future fixed—in which case wouldn't we now be saying that present events depend on the future but not the past? And if there is this alternative way of doing things, how could one way be *right* and the other way *wrong*? How could it be objectively the case, as it seems to be, that we can affect the future but not the past?

This objection might have a second aspect to it, closely related to our present concerns: we might feel that even if the past doesn't depend on the future, it isn't *logically* impossible that it should do so, in the way that this diagnosis would suggest. As David Lewis puts it, this analysis

> gives us more of an asymmetry than we ought to want. ... Careful readers have thought they could make sense of stories of time travel ... ; hard-headed physicists have given serious consideration to tachyons, advanced potentials, and cosmological models with closed timelike curves. Most or all of these phenomena would involve exceptions to the normal asymmetry of counterfactual dependence. It will not do to declare them impossible *a priori*.[2]

The suggested account of counterfactuals thus seems in one sense too weak, in failing to give due credit to a genuine difference between the past and the future; and in another sense too strong, in ruling out backward dependence by fiat. The import of both points is that there is a matter of fact out there, to which the suggested account does not do justice.

What could this matter of fact be, however? The point of our original puzzlement was that we couldn't see any asymmetry in the physics, to explain or underpin the striking asymmetry of dependence. And the main point of chapter 6 was that this first impression is borne out on close examination. The fork asymmetry, which is the only plausible candidate, turns out to be inadequate for the job at hand. But the present objection is that the conventionalist approach takes things too far in the opposite direction. The contrast between 7·1 and 7·2 seems more concrete than the conventionalist suggestion would allow.[3]

In the next section I shall sketch what seems to me to be an adequate response to the first part of the above objection. In other words, I shall show how the conventionalist can explain why *it seems to us* that our ability to affect the future but not the past is an entirely objective matter. At this point one of the useful things about couching the discussion in terms of counterfactuals is that it will enable us to distinguish the convention itself from those features of our "standpoint" which account for the fact that we adopt that convention (and not, say, its temporal inverse).

CONVENTION OBJECTIFIED

The objection is that the conventionalist account would make the temporal asymmetry of dependence less objective than it actually seems to be—a matter of choice, in effect. The appropriate response, I think, is to draw attention to the constraints on this "choice" imposed by the circumstances in which we find ourselves, and over which we have no control. Because these constraints are so familiar we fail to notice them, and mistake the asymmetry they impose for an objective feature of the world. The moral is that things may *seem* more objective than they actually are, when the source of the subjectivity is not an obvious one.

The main constraint turns on the fact that counterfactuals are used in *deliberation,* and that deliberation is a temporally asymmetric process. Considered as structures in spacetime, agents (or "deliberators") are asymmetric, and thus orientable along their temporal axis. The crucial point is that this orientation is not something an agent is able to choose or change. From an agent's point of view it is simply one of the "givens," one of the external constraints within which she or he operates.

In chapter 4 we considered the possibility that elsewhere in the universe, there might be intelligent creatures whose temporal orientation is the opposite of ours; they regard what we call the big bang as lying in the distant future, for example. (Recall that this possibility is a consequence of Gold's hypothesis that entropy would decrease if the universe recontracts.) This possibility illustrates that our temporal orientation may well be a contingent matter, not fixed by the laws of physics. Contingent or not, however, it is not something that we can *change.* We depend on the thermodynamic asymmetry, and for all practical purposes, the second law constrains us in the same way as any genuine law of physics: we can't choose to transport ourselves to a region in which the thermodynamic arrow points the other way, even if such regions actually exist.

Thus for all practical purposes our temporal orientation is fixed. This means that given what we do with counterfactuals in deliberation—more on this in a moment—we have no option but to assess them the way we do; that is, to hold the past rather than the future fixed. Hence it is objectively true that *from our perspective,* we can't affect our past. Unable to adopt any other perspective, however, and failing to notice the relevance and contingency of our temporal orientation, we fail to see that what we have here is a relational truth, a truth about how things are from a perspective, rather than absolute truth about the world. The account thus provides a kind of quasi objectivity. It explains why we think of the asymmetry of dependence as an objective matter.

Again, here's an analogy to illustrate this point. Consider a railway locomotive which cannot travel in reverse, confined to a track without turntables or turning loops. Suppose, for example, that it is confined to the transcontinental line between Perth, Western Australia, and Sydney, New South Wales. Then it is an objective matter which stations the locomotive may visit in the future, though of course a matter which depends on two things: its present location, and its orientation (whether it is facing east or west). The latter factor is relevant only because we have imposed an asymmetric condition on the locomotive's movement, allowing it to move forwards but not backwards. Given this condition, however, from the locomotive's perspective it is an objective matter that some stations are accessible and others are not—and the property of accessibility displays a very marked east-west asymmetry.

My suggestion is that something very similar goes on in the case of deliberation. Given the way in which we are oriented in time, one direction but not the other is accessible to deliberation. Failing to notice the true nature of this constraint, we confuse it for an absolute constraint in the world. If we ourselves had the opposite orientation in time, however, then it would be as if the locomotive were turned around: different parts of reality would seem to be objectively accessible. And if we ourselves were not asymmetric in time, then it would be as if the locomotive could move in either direction. In this case there would be no apparent asymmetry. The last possibility should not be taken too literally, however, for it seems to be an essential feature of deliberation that it is asymmetric.

THE ASYMMETRY OF AGENCY

While the claim that agents are asymmetric, and thus orientable in time, does not seem particularly contentious, it would be useful to have a better understanding as to what this asymmetry involves. I think there are two ways to approach the issue. One would seek to characterize the asymmetry in terms of a formal model of deliberation. Roughly speaking, the goal would be to map the structure of an ideal deliberative process—to map it from an atemporal perspective, laying out the steps *en bloc*—and hence to be able to point to an intrinsic asymmetry along the temporal axis. This seems a plausible project, involving little more than rewriting standard dynamic models of deliberation in an explicitly atemporal key.

The other possible approach to the asymmetry of agency is a phenomeno-logical one, one which appeals to what it *feels like* to be an agent. The story might go something like this. From an objective standpoint, very crudely, an agent is simply a natural system which correlates inputs with outputs. The inputs are environmental data and the outputs are behavior. The details of these correlations vary with the agent's internal state, and this too may vary in response to inputs. The terms "input" and "output" assume a temporal direction, of course, but this is inessential. From an atemporal viewpoint what matters is that events on one temporal side of the box get correlated with events on the other side. It doesn't matter that one side is thought of as earlier and the other later. From a sufficiently detached perspective, then, deliberation appears broadly symmetric in time—an agent is simply a "black box" which mediates some otherwise unlikely correlations of this kind. Certainly the operations of working models may depend on temporal asymmetry, in the way that actual agents require the thermodynamic gradient, but it is possible to characterize what such a system does, at least in these very crude black box terms, without specifying a temporal direction.

If we characterize agency in these terms then it will seem that the essential asymmetry of agency belongs to the internal perspective, to the experience of *being* an agent. From the inside, as it were, we perceive a difference between (what in the standard temporal frame will be described as) the inputs and the outputs. The inputs appear as fixed, or given, while the outputs are what is open, or potentially subject to our control. It is important to stress the subjectivity of this, for the point was that no such difference between inputs and outputs is apparent from the atemporal perspective—an Augustinian god just sees a nexus for a complicated structure of correlations. The difference between the fixity of the inputs from the past and the openness of the outputs to the future is a feature of the experience *from the inside*—a feature of what it *feels like* to be an agent.[4]

The phenomenological account will perhaps appeal more than the earlier formal approach to philosophers who seek to ground folk concepts in folk experience. Do we need to adjudicate between the two approaches? I don't think so. It seems reasonable to expect that they will turn out to be complementary. In effect, the former approach will simply be describing the internal structure of the black boxes on which the latter approach depends. Together they give us two views of the essential asymmetry of agency.

THE ROLE OF COUNTERFACTUALS

Either way, where do counterfactuals fit into the picture? We need to think about the role of counterfactuals in deliberation. An agent normally has the choice of a number of options, and bases her choice not on the desirability of the immediate options themselves, but on that of what might be expected to follow from them. Thus a typical deliberative move is to take what is given or fixed—or rather, in practice, what is known of what is taken to be fixed—and then to add hypothetically one of the available options, and consider what follows, in accordance with known general principles or laws. The temporal orientation of this pattern of reasoning follows that of the agent's perspective. Broadly speaking, what we hold fixed when assessing counterfactuals according to the standard semantics is what presents itself to us as fixed from our perspective as agents. So long as counterfactuals are to maintain their association with deliberation,[5] in other words, our choice of the "hold the past fixed" convention is governed by the contingent but nevertheless unchangeable fact of our orientation as agents. We have no choice in the matter.

To sum up, is it an objective matter that we can affect the future but not (normally) the past? I would like to compare this to the question as to whether it is an objective matter that lemons are sour and not sweet. Once we understand the rudiments of the biological basis of our sense of taste, it is easy to see that we might well have been constructed so that we found lemons sweet instead of sour. Moreover, it seems plausible to say that given differently equipped communities of this kind, there isn't an objective sense in which one is right and the other wrong. Does this mean that it is not really true that lemons are sour? It seems wrong to say that lemons are *really* tasteless, rather than sour. The best option seems to be to say that in this form the question doesn't really make sense from the neutral perspective we adopt when we consider the possibility of differently equipped tasters. From the external perspective it only makes sense to say that lemons taste sweet to group A and sour to group B. The absolute form of speech only makes sense from within, and from this perspective—our perspective—it is simply true that lemons are sour. Whatever we choose to say from the neutral perspective, it should be compatible with the fact that for ordinary purposes it is *obviously* true that lemons are sour.

The situation is similar with asymmetric dependence. We can imagine reverse-oriented creatures who would regard dependence and causation as going, in our terms, from future to past. Moreover we can see that from the atemporal perspective there isn't a fact of the matter as to which orientation gets it right. And yet at the same time we can see that given what we ordinarily mean by the terms, it is straightforwardly true that we can affect the future but not the past. Philosophically speaking, our progress consists in the fact that we can now see that the source of the causal asymmetry lies in us, not in the world. It is the same sort of progress we make in seeing why we needn't treat tastes and color as natural categories.

It seems to me that this account does an excellent job of satisfying the intuitive constraints. It is compatible with the temporal symmetry of microphysics. And although it represents the asymmetry of dependence as anthropocentric, it is not thereby required to be conventional in the problematic sense: it is not something open to us to change by fiat, as it were. The constraint is entirely objective, albeit relative to our contingent orientation. The fact that we normally take it to be objective *simpliciter* is easily explained, however. Normally we look through our internal constraints, not at them. Hence we don't notice the anthropocentricities in what we see. The more basic they are, the harder they are to spot, and temporal orientation is among the most basic of all. We would have to have been gods—Augustinian gods, at that!—to have seen this all along.

COULD THE PAST DEPEND ON THE FUTURE?

The account outlined above explains the *apparent* objectivity of the asymmetry of dependence, and thus meets the charge that the conventionalist strategy makes the asymmetry too weak. However, the second part of the objection was that the conventionalist strategy is also too strong, in making it a terminological matter that we cannot affect the past. As I noted, many philosophers have felt that this is unacceptable. They have argued that even if there is no backward causation, this is a physical rather than a logical matter.

We might be tempted to try to meet this objection by denying the possibility of backward causation (or dependence). After all, it is also a well-known point of view that backward causation, like time travel, would have paradoxical consequences. Why should a conventionalist go to the trouble of saving space for a possibility which is already so contentious? Instead, why not argue that there are already good grounds for thinking that backward causation is impossible, and that the conventionalist account merely confirms this fact?

The immediate trouble with this suggestion is that it is self-defeating. The original argument against the conventionalist was that the case against backward causation does not seem as strong as it should if it rested on a terminological matter. To the extent that the usual objections to backward causation are not convincing, then, this is evidence that its impossibility is not merely terminological. To the extent that they are convincing, on the other hand, no conventionalist confirmation would seem to be necessary.

In the end, however, the best response is to show that the conventionalist approach does leave a loophole for backward dependence.[6] Indeed, in a sense I'll explain, it actually suggests that we might have reason to *expect* a degree of backward dependence in the world. The loophole concerned is closely related to μInnocence and the issues we discussed in chapter 5. And it is directly connected to an important response sometimes offered to the standard "causal paradox" arguments against backward causation and time travel—as we'll see, the same loophole opens both doors, so to speak. Hence we need to begin with a brief account of the standard argument.

ESCAPING THE PARADOXES OF BACKWARD CAUSATION

The paradoxes of time travel are familiar from many films and science fiction stories. The hero travels into her own past, and takes steps to ensure that she will not exist at the time she left. She kills her young self, introduces her mother-to-be to contraception, or something of the kind. The results are paradoxical: the story tells us that something is both true and not true. It is like being told that this is the best of times and the worst of times, except that we are offered an account of how this contradictory state of affairs came to pass. This has the literary advantage of separating the beginning of the story from the end, and the logical advantage of allowing us to conclude that something in the offered account is false. In physical rather than literary mode we can thus argue against the possibility of time travel: if there were time travel we could design an experiment with contradictory results; *ergo*, there is no time travel.[7] And as for time travel, so for backward causation. Causing one's young self to have been pushed under a bus is just as suicidal as traveling in time to do the deed oneself.

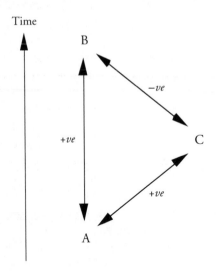

Figure 7.2. The structure of the bilking argument.

As I noted in chapter 5, philosophers call this the bilking argument. To put the argument a little more formally, suppose that someone claims that we can bring about an earlier event of type A by means of a later event of type B (see Figure 7.2). At the very least, this seems to involve the claim that there is (or would be) a reliable positive correlation between B and A on occasions on which we have the option of producing B—that B makes A more likely, as we might ordinarily put it. The bilking argument claims to show that no such correlation could be reliable. Consider Figure 7.2: The claimed positive correlation between A and B could not continue to hold in the presence of a device which first detected whether A had occurred, thus establishing a positive correlation between A and some third event C; and then ensured that B occurs if and only C does not occur, giving a negative correlation between B and C. The pattern of correlations shown in the diagram is logically impossible, so that in the presence of those on the right-hand side, the direct correlation between A and B must fail to hold. In other words, the claimed A-B correlation is "bilked" by the combination of correlations between A and C and between not-C and B.

The argument is thus that the possibility of bilking shows that there cannot be a reliable (or "lawlike") correlation between past states and *independently determined* future states. The qualification is crucial, of course. The bilking argument does not exclude normal cases in which an earlier event causes a later event. In these cases the lawlike claim is a circumscribed one: B follows A *in appropriate circumstances*. Bilking destroys those circumstances.

Here is a commonplace example to illustrate this point: Drinking cocktails is positively correlated with social relaxation and morning-after headaches. Couldn't we bilk the latter correlation by establishing a negative correlation between relaxation and morning-after headaches? Of course: we would simply have to arrange that nervous subjects took headache-inducing medication, and relaxed subjects took headache-preventing medication. Equally obviously, however, this possibility doesn't defeat the claim that there is a reliable correlation between drinking and hangovers. It is simply that like all (or almost all) reliable correlations, it only holds in appropriate circumstances. The intervention we imagined (headache-inducing medication, etc.) would be incompatible with these circumstances, and this is why the bilking argument doesn't count against familiar correlations of this kind.

In the abnormal case we are interested in, however, the correlation A-B is claimed to hold even when B is independently sensitive to external factors—even in the presence, in particular, of whatever it takes to establish the negative correlation between B and C. In this case the bilking argument may seem to be on solid ground, but there is still the role of the third side of the triangle to consider. It might be suggested that the mechanism responsible for the A-C correlation inevitably disturbs the circumstances required for the A-B correlation.

In the usual presentation of the bilking argument, the mechanism responsible for the A-C correlation is simply some sort of observational procedure to determine whether A has occurred. (C just corresponds to a positive result.) It is usually taken for granted that such an observational procedure will not disturb the circumstances in such a way as to leave the loophole in the argument that we naturally exploit in the time-reverse case. That is, it is assumed that the presence of such an observational procedure could not provide a plausible explanation for the failure of the contentious A-B correlation. With this assumption in place, we reach the conclusion that the Oxford philosopher Michael Dummett reached some thirty years ago. In Dummett's terms, the conclusion is that claims to affect the past are coherent, so long as it is not possible to find out whether the claimed earlier effect has occurred, before the occurrence of the alleged later cause.[8]

Dummett's own illustration involves a tribe who conduct a ritual to try to ensure the success of a hunt which has already taken place, but from which the hunters have not yet returned. He argues that there is no inconsistency in the tribe's beliefs, so long as they regard it as impossible to find out whether the hunt has been successful, before they perform the ritual in question. This ensures that from their point of view, the bilking experiment appears impossible to perform. (They might be wrong about this, of course, but the issue is whether a belief in backward causation is *coherent,* not whether it is *true.*)

Dummett's formulation of the condition under which a belief in backward causation is coherent is a little misleading, however. Because it takes for granted that the means by which we find out about past events don't disturb things in any relevant way, it hampers attempts to think about abnormal past-directed correlations by analogy with the normal future-directed case. As we have seen, what defeats the bilking argument in the latter case is not the nonexistence of the correlations required to close the triangle, but the fact that the mechanisms of such correlations disturb the circumstances required for the correlations under challenge. So a perfect parallel would be a case in which the past-directed bilking argument was defeated not by the nonexistence of a means to observe the relevant past events—as Dummett assumes—but by the fact that the mechanism of such observations inevitably disturbed the system in such a way that the claimed correlation failed to hold. In Dummett's terms, for example, the parallel case would be that in which the ritual was believed to be effective only under conditions which would be defeated by any attempt to discover the outcome of the hunt in advance. (We'll come to an example with this feature in a moment.)

Dummett's strategy is not the only defense of the coherency of backward causation against the bilking argument. There is a quite separate tradition in the literature which argues that the bilking argument is inconclusive even in the presence of normal epistemological "access" to past events.[9] Where available, however, Dummett's is by far the strongest defense. As we shall see later in the book, it is also the one of particular relevance to contemporary physics. In order to appreciate this point, however, we need first to appreciate a connection between the Dummettian loophole in the bilking argument and the one that enables our conventionalist approach to make sense of backward dependence.

THE PAST UNLOCKED

We are now in a position to respond to the objection that the conventionalist view makes it a matter of definition that earlier events cannot be counterfactually dependent on later events. The response turns on the fact that although it is a matter of definition that we can't affect what we *know about* at the time of deliberation, it is certainly not a matter of definition that what we can know about is whatever lies in our past. Indeed, it is an interesting question why we tend to assume that this is the case: that is, why we think that whatever lies in the past is knowable as we deliberate, at least in principle. Even if experience teaches us that whatever we know about via memory and the senses lies in the past, this does not imply that anything that lies in the past is something that might in principle be known about, and hence something accessible in the process of deliberation.

In fact, it seems that the relationship between temporal location and epistemological accessibility is not only contingent (in both directions), but rather underdetermined by our actual experience. For all that our limited experience tells us, there might be some of the past to which we do not have access, and perhaps some of the future to which we do. The epistemological boundaries seem to be an empirical matter, to an extent that ordinary practice tends to overlook.

Here the point connects with the discussion in the previous section. We saw that it is possible to avoid the bilking argument so long as one confines oneself to the claim that one can affect bits of the past which are epistemologically inaccessible at the time that one acts. To put the point in terms of a concrete example, suppose it were suggested that the present state ϕ of our incoming photon depended on the setting of the local polarizer in its future, as well as on that of the distant polarizer in its past. In this case, as we saw just now, the bilking argument would require that we measure ϕ before the photon arrives, and thereby set the future polarizer to conflict with the claimed correlation. So if the only way to measure ϕ is to pass the photon through a polarizer, and the claim in question is simply that the value of ϕ depends on the orientation of the *next* polarizer it encounters, bilking becomes impossible. Interposing another measurement would mean that the *next* measurement is not the one it would have been otherwise, so that the conditions of the claimed dependence would no longer obtain. According to the claim under test, such a measurement cannot be expected to reveal the value that ϕ *would have had* if the measurement had not been made.

By slipping into talk of counterfactuals and backward dependence here, I have tried to illustrate that this admission of limited backward dependence is not as alien as might be imagined. It seems in particular that our use of counterfactuals is already somewhat more flexible than the model we have been using suggests—already sufficiently flexible to handle the kind of cases that Dummett's loophole admits, in fact. What we seem to do in such cases is just what Dummett would have us do, in effect: we assess counterfactuals not by holding fixed everything prior to the time of the antecedent condition, but by holding fixed what we have access to at that time (without disturbing the background conditions of the claimed correlation).

This shows how the conventionalist can make sense of the possibility of backward dependence, and thus meet the second part of the objectivist's challenge. Note that this response does not require that the ordinary use of counterfactuals is already unambiguously committed to this possibility, in the sense that it is already configured to take advantage of Dummett's loophole. If that were so the assumption that the past does not depend on the future would surely be less deeply ingrained than it is in scientific theorizing. It seems to me more accurate to say that there is a systematic indeterminacy in our ordinary notions of causal and counterfactual dependence. Ordinary usage does not clearly distinguish two possible conventions for assessing counterfactuals: a convention according to which the entire past is held fixed, and a convention according to which only the *accessible* past is held fixed. If we follow the former convention then it is a matter of definition that the state of the incoming photon does not depend counterfactually on the local polarizer setting; whereas if we follow the latter convention then it may do so. But this difference reflects an issue about how we should use the terms involved, rather than a disagreement about the objective facts of the matter.

This may seem like cheating, however, for in what sense does the conventionalist strategy actually make sense of the possibility of the past depending on the future? I have said that it only does so if the convention governing counterfactuals in understood in the right way. Doesn't this amount to little more than the claim that the past may depend on the future if we define "dependence" so as to allow it?

Two responses to this objection: First, the proposal has not redefined "dependence," but simply pointed to an ambiguity in its existing meaning. I have argued that under one resolution of this ambiguity, well in keeping with our ordinary intuitions, the past might indeed be said to depend on the future. This possibility seems to provide as much as we need by way of a response to the original objection, and doesn't involve any illegitimate redefinition of terms.

Second, and more important, the above discussion shows that the conventionalist can make sense of an objective issue about the correlational structure of the world, an issue which is not dependent upon how we choose to resolve the ambiguity just mentioned. Indeed, by dissecting out the conventional ingredient in our ordinary view, the conventionalist is able to call attention to the objective remainder in a way that other accounts find extremely difficult. And here we come to the big payoff. It turns out that there is a possible "world structure" which has been almost entirely ignored by contemporary physics—partly, I think, because it has been assumed that the concepts on which its visibility depends were not really a matter for physics, being subjective (or "metaphysical," in the disparaging sense that physicists sometimes give to this term). In some respects this intuition is correct. However, it turns out that when we prune away the subjective and metaphysical thorns, a physical kernel remains—a kernel which turns out, as we'll see, to be profoundly relevant to some of the deepest problems in contemporary physics.

ADVANCED ACTION: ITS OBJECTIVE CORE

It will be helpful to have a name for the objective core just described. With an eye to its relevance to quantum mechanics I shall borrow a term in use in physics, and call it *advanced action*. What advanced action actually involves will become clearer as we proceed. Eventually, we'll see that it is closely related to the possibility of abandoning μInnocence. However, a useful way to get a preliminary fix on the notion is to triangulate in terms of our two possible accounts of the meaning of counterfactual claims. We have outlined two possible conventions for counterfactuals. According to one convention, we hold fixed everything prior to the time of the antecedent event. (In the case of 7.3, for example—"If X had happened at time t_1 then Y would have happened at time t_2"—this means everything prior to t_1.) According to the other convention, we hold fixed merely what we have access to at that time.

Let's think about what advocates of these two conventions might agree about, in circumstances in which they disagreed about the truth of particular counterfactuals. In the photon case, for example, they might agree that there is a one-to-one correlation between some feature of the state of the incoming photon and the setting of the later polarizer.[10] Putting it in the more intuitive terms we used in chapter 5, they might agree that the photon "knows" the setting of the polarizer, before it reaches it; at any rate, they both accept that there is a strong correlation between the setting of the polarizer and the earlier condition of the photon.

Despite agreeing on this, the two views may disagree as to whether if the setting of the polarizer had been different, the prior state of the photon would also have been different. According to the "hold the past fixed" convention, this counterfactual is false by definition. According to the "hold what is accessible fixed" convention, it may well be true. However, both sides will be able to recognize that this isn't a genuine disagreement between them, but merely a verbal one. It simply stems from the fact that they mean different things by their counterfactual claims.

Thus by separating the issue of the correlational structure of the world from that of the appropriate convention for counterfactuals, we seem to be able to insulate an objective issue, a question concerning the structure of the world, from the effects of folk intuitions about what might depend on what. Somewhat indirectly, then, we can say that a world with advanced action would be a world in which the two conventions disagreed—a world like the one in which the photon knows in advance about the setting of the polarizer, in which the two conventions give different answers and we have to make a choice.

The point of approaching things in this indirect way is to show that the issue isn't entirely a matter of choice between conventions; that when we divide through by the conventions, there is a real physical remainder. Indeed, the argument doesn't even require that all parties concede the coherence of both the relevant conventions for counterfactuals. Faced with opponents who claim to find backward dependence incoherent, for example, we could simply concede counterfactual practice—give them their convention—and fall back to the underlying objective issue.

It would be helpful to have a more direct way of thinking about this objective issue. In a moment I want to show that it corresponds to the possibility we thought about in a more intuitive way in chapter 5, when we considered giving up μInnocence. First, however, I want to say something more about the choice between our two conventions for counterfactuals. In adopting an even-handed approach just now, our aim was to keep our eyes on the issue that really matters for physics. This strategic consideration does not prevent us from taking sides, eventually, on the question as to which convention would be the more appropriate, in a world with correlations of the imagined kind. I think there are good reasons for preferring the weaker convention ("hold fixed the *accessible* past"), which become evident if we consider what the stronger convention would have us say about the imagined case involving photons and polarizers. The next section is devoted to this; it is a little technical, and may be skipped without loss of continuity.

COUNTERFACTUALS: WHAT SHOULD WE FIX?

To simplify as much as possible, suppose that all parties agree that there is a strict correlation between a particular polarizer setting S and a particular earlier state ϕ_s of the incoming photon. If counterfactuals are assessed according to the stronger convention—"hold the entire past fixed"—it cannot be true both that

(7.4) If we were to bring about S, it would be the case that ϕ_s

and that

(7.5) If we were to bring about not-S, it would be the case that not-ϕ_s.

For if we hold fixed events before S, the consequent (i.e., second clause) of one of these counterfactuals is bound to be false, even under the supposition in question.

Suppose for the sake of argument that the incoming photon is not in state ϕ_s, so that it is 7·4 whose consequent is false (under the assumption that the past is held fixed). The two possibilities seem to be to regard 7·4 as false or to regard it as somehow meaningless, or otherwise inappropriate. To go the first way is to say that the agreed correlation between S and ϕ_s does not support counterfactuals. To go the second way seems to be to say that means-end reasoning breaks down here—that it doesn't make sense to suppose we might do S. Neither course seems particularly satisfactory. We are supposing that all parties acknowledge that as a matter of fact the correlation does always obtain between S and ϕ_s, even on whatever future occasions there might happen to be on which we bring about S. Outcomes of actual actions thus assured, it is hard to see how the refusal to acknowledge the corresponding counterfactuals could seem anything but willful—so long, at any rate, as we claim the ability to bring about S at will. Denying free will seems to be an alternative, but in this case it should be noted that the phenomenology isn't going to be any different. So there is nothing to stop us from going through the motions of deliberating, as it were. Within this scheme of quasi deliberation we'll encounter quasi counterfactuals, and the question as to how these should be assessed will arise again. Hold fixed the past, and the same difficulties arise all over again. Hold fixed merely what is accessible, on the other hand, and it will be difficult to see why this course was not chosen from the beginning.

Thus I think that if we found ourselves in a world in which the notions of the past and the epistemologically accessible came apart in this way—a world in which it was therefore important to resolve the ambiguity of current counterfactual usage—a resolution in favor of holding fixed merely what is accessible would be the more satisfactory. For present purposes, however, the important point is that this issue about how we should use counterfactuals is quite independent of the empirical issue as to whether the world has the correlational structure that would require us to make the choice.

ADVANCED ACTION AND µINNOCENCE

Let's review the story so far. We began the chapter with the temporal asymmetry of counterfactual dependence, in the form of the contrast between 7·1 and 7·2. I argued that this asymmetry reflects a conventional asymmetry in the way we use counterfactual claims, rather than an objective asymmetry in the world. I pointed out that we can explain its apparent objectivity in terms of the fact that the convention isn't a matter of choice: given our own temporal orientation, and what we want to do with counterfactuals, we don't have the option to use the reverse convention.

More surprisingly, it has turned out that this account leaves open the possibility of exceptional cases, in which the past could properly be said to depend on the future. True, it leaves open this possibility not in the direct sense that current usage unambiguously admits it, but in the sense that in conceivable physical circumstances the most natural way to clarify and disambiguate current usage would be such as to recognize such backward influence. But the underlying physical circumstances themselves do not depend on our choice: it would be an objective feature of the world that it required us to make a choice. I referred to this objective possibility as that of "advanced action."

Thus far, however, I have only characterized advanced action in a very indirect way. In effect, I have marked out an objective possibility in terms of two subjective notions. I want now to try to provide a more direct and intuitive picture of what this possibility involves, by connecting advanced action to μInnocence, and the other issues we discussed in chapter 5. In the process, I want to show in what sense T-symmetry counts in favor of advanced action, as it seemed to count in favor of abandoning μInnocence.

The importance of symmetry in this context lies in the fact that any argument for advanced action seems likely to be indirect. After all, the space we have found for this possibility lies in the gap between the past and the *accessible* past, where these notions come apart. The past which might coherently be taken to depend on our present actions is the *inaccessible* past—that bit of the past which we cannot simply "observe," before we act to bring it about. Where observation operates in what we assume to be the normal way—that is, where the entire past is accessible, at least in principle—the two conventions for counterfactuals come to the same thing, Dummett's loophole is not available, and the bilking argument goes through (modulo the concerns mentioned in note 7, at any rate).

Thus if a case is to be made for advanced action—for the view that we actually live in a world in which the two conventions for counterfactuals do not amount to the same thing—it will need to be indirect. The argument will have to rely on nonobservational considerations: factors such as simplicity, elegance, and symmetry, for example. This doesn't mean that the issue is nonempirical, or metaphysical in the disparaging sense. Factors of this sort often play an important role in science. Later in the book some of the advantages of advanced action will emerge indirectly, from the tribulations of alternative approaches to quantum mechanics. At least one advantage might be more immediate, however: it is a tempting thought that there should be a symmetry argument in favor of the advanced action view.

If there is a symmetry argument for advanced action, we might expect it to be closely related to the issues we discussed in chapter 5. I argued there that symmetry counts against μInnocence: once we see that μInnocence does not have an observational basis, we see that it can only be an independent theoretical assumption, in conflict with T-symmetry. However, we saw that abandoning μInnocence seemed likely to commit us to a kind of backward causation. Is this the same possibility that I am now calling advanced action? If so, the same symmetry argument might be expected to work in both cases.

IS μINNOCENCE MERELY CONVENTIONAL?

We need to be careful here, however. As I noted at the end of chapter 5, physicists attracted to the idea that our causal intuitions are at least partly subjective have often taken this to imply that for symmetry reasons, a non-subjective view would recognize causation in both directions on an equal footing. (A drastic form of this view is the eliminativist option, mentioned in chapter 6, which dispenses with causation and dependence in physics altogether.) Far from admitting advanced action, then, this approach tends to reject even the ordinary kind of "retarded" or "forward" causation. After all, if the asymmetry of causation is perspectival, what room is there for the view that certain causal processes go backwards rather than forwards? Thus if we want to advocate both advanced action and a perspectival view of causal asymmetry, we must be very careful not to throw the baby out with the bath water.[11]

Another way to make this point is to note that, at least at first sight, the perspectival approach to causal asymmetry seems to explain the asymmetry of μInnocence, *without invoking any objective asymmetry in the world.* In other words, it seems to explain our intuitions about μInnocence in terms of an asymmetry *in us,* not an asymmetry in the physical systems to which we take it to apply. How? Well, consider a simple and clearly symmetric physical example, such as the collision of two (frictionless) Newtonian billiard balls. If we know the combined momentum of the balls before they collide, then conservation of momentum ensures that we also know it after the collision. After the collision, then, the momentum of the balls is correlated, in the sense that by measuring the momentum of one ball, we can determine that of the other.

In one important sense, however, there is nothing asymmetric about this. After all, if we know the combined momentum *after* the collision (which is the same thing as knowing the momentum before the collision, of course, thanks to conservation of momentum), then we know that the momentum of the balls *before* the collision is correlated in just the same way. (The fact that there is nothing intrinsically asymmetric in cases like this was the point of Figure 6.2.)

All the same, the asymmetric convention we apply to counterfactuals seems to explain why we make asymmetrical judgments about this kind of case. For example, we say that given that the two balls X and Y collide at time t, their momenta become correlated in the sense that

(7·6)   If the momentum of X after t had been larger, that of Y would have been correspondingly smaller, and vice versa.

In apparent contrast, we don't accept that given that X and Y collide at time t, their momenta are correlated in the sense that

(7·7)   If the momentum of X *before* t had been larger, that of Y would have been correspondingly smaller, and vice versa.

As we have seen, the difference in our attitude to these claims seems to arise from a conventional asymmetry in the way we assess counterfactuals in general. In the case of 7·6 we seem to "hold fixed" the total momentum *before* the time t in question, and appeal to conservation of momentum. In the case of 7·7, however, we don't do the same thing in reverse: we don't hold fixed the total momentum *after* t. The physical symmetry of the situation ensures that if we were to do so, we would come up with the same answer in both cases. (This is simply to reiterate the lesson of 7·1 and 7·2, of course: because 7·1 and 7·2 are not genuinely the temporal inverses of one another, their difference in truth value does not require a temporal asymmetry "in the world.")

However, 7·6 and 7·7 seem to illustrate what we have in mind when we say that physical systems are correlated after they interact but not before they do so. On the face of it, then, we seem to have explained the asymmetry of μInnocence, *without appealing to any objective asymmetry in the physical systems themselves.* This would be a devastating objection to the argument of chapter 5—to the idea that μInnocence embodies an objective asymmetry, whose significance physics has overlooked.

At this stage, then, it looks as though the attempt to find a symmetry argument in favor of advanced action has backfired in a spectacular way. In trying to link advanced action to μInnocence, we seem to have discovered not that there is a symmetry argument in favor of advanced action, but that there isn't any such argument against the intuitive asymmetry of μInnocence.

Fortunately for the project of the book, however, first impressions turn out to be misleading at this point. It turns out that there is another asymmetry related to μInnocence, which is not reducible to the conventional asymmetry of counterfactual dependence. This additional asymmetry is more selective than that of counterfactual dependence. It is present in our intuitive picture of the behavior of photons and polarizers, but not of billiard balls. By distinguishing it from the conventional asymmetry, we will be able to get a better idea of the sense in which T-symmetry counts in favor of the advanced action proposal.

WHY CAN'T A PHOTON BE MORE LIKE A BILLIARD BALL?

First, then, to the objective asymmetry involved in our intuitive commitment to *μ*Innocence. I want to show that there is an asymmetric pattern of correlations—a pattern *not* exhibited by symmetric processes such as our collision of Newtonian billiard balls—which we nevertheless regard as natural and intuitively plausible.

As in earlier chapters, the easiest way to make the target asymmetry stand out is to think about how the processes in which we take it to be involved would look in reverse. Imagine, then, that we are being shown video clips of microscopic interactions—not real photographic images, but diagrammatic representations—and that we are being asked to answer two questions:

- Is the process shown of a kind which occurs naturally in the real world?

- Is the reverse process—what we see if the video is played in reverse—of a kind which occurs naturally in the real world?

If the video clips show interactions between a few Newtonian particles, or the frictionless billiard balls we considered a moment ago, then it is easy to see that both parts of this little quiz have the same answer. If a given process looks like a natural or realistic Newtonian interaction, then so does the reverse process. In effect, this is just what the T-symmetry of Newtonian mechanics amounts to. (Remember that we are dealing with individual microscopic processes, not with the statistical behavior of large numbers of individual constituents, so the familiar macroscopic asymmetries will not show up.)

By contrast, consider a video clip which depicts a series of interactions between incoming photons and a polarizer. In each case, a photon passes through the polarizer, and *afterwards* has a state which exactly reflects the orientation of the polarizer. The photons are not depicted as being correlated with the polarizer *before* they encounter it, however. This behavior looks "natural," or realistic: most of us would answer "yes" to the first question. The reverse behavior strikes us as unrealistic, however, so we are inclined to say "no" to the second question.

Thus there is an intuitively natural asymmetry in the way we think about the world, which can't be explained by the conventional asymmetry of counterfactual dependence. Why not? Simply because, as we saw, the counterfactual asymmetry shows up in the billiard ball case, as well as the photon case—which means that it cannot be the basis of our asymmetric intuitions about the quiz just described, which apply to the photons but not to the billiard balls. (Note that it is irrelevant here whether the video shows "photons as they really are," or even whether this is possible. The point of the example lies in what it tells us about our ordinary intuitions. A completely fictional example would have served our purposes almost as well.)

In effect, then, we have discovered that our ordinary intuitions about μInnocence actually embrace two quite separate principles. In cases like that of the billiard balls, what we mean by the claim that incoming systems are independent is simply what is given to us by the conventional asymmetry of counterfactual dependence. This doesn't involve any temporal asymmetry in the objects themselves. In cases like that of the photons, however, our conception of what counts as an intuitively plausible model of the world is guided by another time-asymmetric principle. We regard as natural the kind of asymmetric pattern of correlations which enables us to distinguish between the forward and reverse videos of the photon interactions. In this noncounterfactual sense, we expect physical systems to be correlated after but not before they interact. This second principle remains in conflict with T-symmetry, even when the asymmetry of the first is explained away in conventionalist terms.

Up to this point, then, our use of the term μInnocence has been ambiguous. Let us restrict it to the second kind of asymmetry, so that Newtonian billiard balls (even microscopic ones) do not exhibit μInnocence. What is the connection between μInnocence in this restricted sense and advanced action? In particular, does abandoning μInnocence automatically commit us to advanced action? If so, then the symmetry argument against μInnocence is also a symmetry argument in favor of advanced action.

However, the billiard ball examples seem to suggest that abandoning μInnocence does not necessarily involve advanced action. As we have just seen, these cases do not exhibit μInnocence (at least so long as we ignore friction and the like)—and yet isn't it obvious that they involve no advanced action? When two billiard balls collide, isn't it simply absurd to suggest that the properties of one might depend on the properties of the other—that we could influence one (before the collision) by altering the properties of the other? If so, then symmetry doesn't seem to commit us to advanced action.

We have to be careful, however. If we were asked to defend the claim that we cannot influence one billiard ball by manipulating the other, we might well fall back on the bilking argument, to try to show that the suggestion leads to the familiar causal paradoxes. Of course, there doesn't seem to be any objection to the bilking argument's accessibility requirement in a case like this: we seem to be justified in assuming that we could observe the

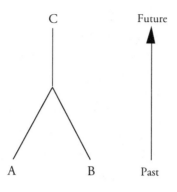

Figure 7.3. The three-legged diagram.

relevant properties of the remote billiard ball, before we decide what to do with the local one. However, our ability to do this depends on the fact that in practice, the billiard balls are not isolated physical systems. Each billiard ball is involved in a vast array of interactions with other elements of the world, some of which we exploit when we "observe" or "measure" the ball in question. When all these interactions are taken into account, it is far from clear that the billiard balls do constitute a symmetric system—a system without μInnocence in the restricted sense. So it isn't at all clear that simple Newtonian systems provide a counterexample to the claim that symmetry requires advanced action.

SYMMETRY AND ADVANCED ACTION I

Let us think about the problem in rather more abstract terms. We are interested in cases in which incoming physical systems interact for the first time. We may represent such an interaction by the three-legged diagram in Figure 7.3. The two legs to the past represent the incoming systems, while the leg to the future represents everything that happens to the combined system after the interaction. The question we want to think about is this: what happens if we introduce a change on one of the incoming legs? What effect does this have on the other two legs?

Usually we assume that the effect will be confined to the future leg. In the billiard ball case, for example, the initial change might correspond to an increase in the momentum of ball A. We expect the effect of this, in light of the principle of conservation of momentum, to be an increase in the total momentum of the combined system C. However, it is also easy to see that this result is not dictated by conservation of momentum alone. The conservation law would be equally happy if the increase in the momentum of A was matched by a decrease in the momentum of B. And there is a whole range of intermediate cases, in which part of the increase shows up at C, and part is matched by a decrease at B. Why then do we take for granted what seems to be an extreme point on this range of possibilities, namely that the increase all shows up at C?

We have seen that as long as we assume that we have access to the relevant properties on path B, at the time of any change on A, the bilking argument seems to force us to absorb the entire change on C. In this case, any other possibility threatens contradictions.[12] Without such access, however, what principle forces us to make this choice—except, of course, some version of μInnocence itself?

In terms of these three-legged diagrams, we can think of advanced action as the hypothesis that changes on one of the past legs are not absorbed entirely on the future leg. We haven't yet shown that symmetry commits us to this possibility, but we have seen that in rejecting the asymmetric principle μInnocence, we reject the only apparent objection to this possibility.

Thus in the billiard ball case, advanced action seems to be allowed but not mandated by the kind of temporal symmetry involved in giving up μInnocence. Advanced action would only be required, rather than merely possible, if we had some reason to think that a change in A was not fully absorbed on C—in other words, that the future combined system would remain wholly or partly unchanged, despite a change in A. However, we don't seem to find any such reason in the billiard ball case.

The case of the photon is rather different. Let us now think of A in Figure 7.3 as the history of the polarizer, B as that of the incoming photon, and C as the future in which the photon has passed through the polarizer. (C thus contains two elements, or subfutures: that of the photon after the interaction, and that of the polarizer.) Roughly, to say that changes in A are absorbed only in C is to say that A and B vary independently of one another, in real cases with futures like C. Normally, this sort of independent variation strikes us as intuitively plausible. The key intuition is μInnocence, in effect, and as our video example revealed, it is strongly asymmetric. The time-reversed cases, in which the state of a photon and a polarizer are uncorrelated after they interact, do not seem at all plausible.

Restoring T-symmetry would be a matter of rejecting μInnocence, and allowing correlations between the photon and the polarizer in both temporal directions. In this case, then, symmetry seems to support advanced action in the strong sense that if correlations toward the past are simply allowed to match the correlations which we take to hold toward the future, advanced action is the result.[13]

SYMMETRY AND ADVANCED ACTION II

Thus in the photon case, it seems that we reject advanced action at the cost of endorsing an objective temporal asymmetry—an asymmetry which unlike that of counterfactual dependence, cannot be explained away as a mere manifestation of our own temporal perspective. We approached this conclusion from the bottom up, as it were, by thinking about μInnocence, and what its rejection would imply in the photon case. I now want to show that we can get to a parallel conclusion from the opposite direction. If we think simply about counterfactuals, we find that our intuitions about the photon case reveal a temporal asymmetry, not explained by the general asymmetry of counterfactual dependence.

Consider again the original photon case, in which the photon passes through two polarizers before it reaches us. The intuitive view is that the state ϕ of the photon in the interval between the polarizers does not depend on the orientation of the future polarizer, which the photon has not yet reached. If we assume—reasonably, as we did earlier—that what is accessible to us at this stage is at most the state of the photon in the region prior to the past polarizer, then an advocate of the "hold fixed what is accessible" convention for counterfactuals will read this independence assumption as follows:

(7.8)   With the history prior to the past polarizer held fixed, changes in the setting of the future polarizer do not imply changes in the value of ϕ in the region between the polarizers.

In order to determine whether the intuitive view involves a temporal asymmetry, we need to ask whether it also endorses the temporal inverse of this assumption, which is:

(7.9)   With the course of events after the photon passes the *future* polarizer held fixed, changes in the setting of the *past* polarizer do not imply changes in the value of ϕ in the region between the polarizers.

There is little intuitive appeal in this latter proposition, however. The easiest way to see this is to imagine more familiar cases in which we talk about alternative histories. For example, suppose that we have an artifact, known to be a year old, which could have been manufactured by one of two processes. Given its present condition, does our view about its condition, say, six months ago depend on our view about its origins? Obviously it might well do so. In the one case, for example, the distinctive patina on the surface of the object might have been a product of the manufacturing process itself; in the other case it might have been acquired gradually as the object aged. So there is no reason to expect the condition of the object in the intervening period to be independent of what happened to it in the past.

In endorsing 7·8 but not 7·9, then, the orthodox view seems to be committed to a genuine asymmetry—the sort of asymmetry which is *not* needed to account for the contrast between 7·1 and 7·2. The advanced action view would avoid this asymmetry, by rejecting 7·8 as well as 7·9.

This asymmetry does not rest on some special peculiarity of the photon case. As I noted at the beginning of chapter 6, there is an objective asymmetry embodied in the standard interpretation of quantum mechanics, according to which the wave function is localized *after* but not *before* a measurement interaction. This provides countless examples of the kind of asymmetry just described. In the photon case I have been relying more on naive physical intuitions than on the quantum mechanical account of interactions between photons and polarizers. The standard quantum mechanical account fits the naive picture, however: if we know the quantum state function of the photon before it reaches the distant polarizer, and vary our assumptions about the setting of the local polarizer, this has no bearing on the state of the photon between the polarizers. Whereas if we are told the state of the photon after it passes the local polarizer, and are asked to consider varying assumptions about the setting of the distant polarizer, we derive varying conclusions about its state in the region between the polarizers. The standard account gives us 7·8 but not 7·9, in other words. As we shall see in chapter 8, the same is true quite generally: the orthodox view of quantum mechanics is thoroughly asymmetric in this way.

An advanced action view restores symmetry by rejecting 7·8 as well as 7·9. Symmetry thus seems to provide a powerful prima facie argument in favor of the advanced action view. This may seem implausible—how could such a simple argument have been overlooked, after all? Perhaps in part because we tend to think about these issues in terms of counterfactuals, and tend to think about counterfactuals in terms of the simpler "hold fixed the past" convention. In this case the assumption that the past does not depend on the future amounts to nothing more than the contrast between 7·1 and 7·2, which is rightly assumed to involve no problematic temporal asymmetry. In order to notice the problem we need to prise various things apart. In particular, if we want to use counterfactuals then we need to recognize the possibility of using the "hold fixed the *accessible* past" convention.

Counterfactuals are not essential, however. As we saw in chapter 5, and confirmed in the previous section, we can quite well characterize the issue in more basic terms.[14] But in this case a second factor tends to obscure this symmetry argument for advanced action, I think. It is our tendency to take for granted the availability of "classical" measurement, which combines with the bilking argument to enforce the temporal asymmetry that the advanced action view rejects. Given classical measurement, the bilking argument gives us a reason to reject 7·8. In other words, classical measurement appears to underwrite the objective asymmetry which is reflected in the contrast between 7·8 and 7·9. Perhaps the right thing all along would have been to turn things around. As I suggested in chapter 5, perhaps we should have already been suspicious of classical measurement, because it conflicts with temporal symmetry in this way.

At any rate, where the world does not provide classical nondisturbing measurement, then symmetry seems to be a possibility. It is here that what is past and what is accessible come apart, and Dummett's strategy allows us to unlock a little of the past. The two conventions for counterfactuals differ in theory, and may disagree in practice. I characterized advanced action in terms of this possibility. Worlds with advanced action are worlds in which the two conventions do disagree. It seems that only a world of this kind can avoid objective temporal asymmetry of the kind described in this section.

TAXONOMY AND T-SYMMETRY

Here's another way to think about these issues. Ordinarily we take it for granted that what happens in the future cannot make a difference to the nature of a physical system *now*. In other words, we take it for granted that physics cannot put systems in different categories—treat them as being of different *kinds*—simply on the grounds that they have different futures. Taxonomy is blind to the future, as we might say. Clearly, this intuition is very close to that of μInnocence.

It is easy to see why taxonomy *as a human practice* should involve this temporal asymmetry. Roughly speaking, we group objects into categories according to what we can find out about them, and what we can find out concerns their past. In this respect, our taxonomy reflects our own temporal asymmetry as *knowers*, in much the same way that our use of causal notions reflects our own temporal asymmetry as *agents*. But once we notice this, why should we continue to think that the taxonomy of the objects in themselves is governed by this asymmetry? Why shouldn't we suppose, instead, that the real taxonomy is from our standpoint teleological: that is, that there are real distinctions between objects which turn on what is to happen to them in the future?

As long as we assume that the entire past is accessible to knowledge, at least in principle, then the bilking argument stands in the way of such a time-symmetric taxonomy. After all, this assumption seems to imply that if there were real forward-looking distinctions of this kind, then in principle we might discover them, and alter the future in the familiar paradoxical ways. But what this shows, once again, is that symmetry should already have made us suspicious of the assumption that everything in the past is accessible to knowledge. For we surely have no reason to think that the bare taxonomy of the world is time-asymmetric in such a striking way.[15]

BACKWARD CAUSATION: NOT FORWARD CAUSATION BACKWARDS

What would a symmetric world be like, on this view? We have seen that it would involve a kind of advanced influence, or backward causation, but it is very important to appreciate that this would not be simply the temporal inverse of ordinary forward causation. The easiest way to appreciate this point is to consider the kind of diagram we used in Figure 7.3. Let us make the diagram four-legged, as in Figure 7.4, so as to be able to represent separately an agent (depicted by the short left-hand legs) and an external system (the longer right-hand legs), both before and after they interact. Thus in each of the four individual diagrams in Figure 7.4, the left-hand legs represent the agent, and the right-hand legs represent the object system. The diagram at the top left represents what we think of as ordinary "forward" causation, in which the effects of the agent's intervention (represented as the thick shaded arrow) show up on the paths of the agent and the object system *after* the interaction; that is, on the two future legs of the diagram.

Advanced action is represented by the top right-hand diagram in Figure 7.4. Here the effects of the agent's intervention (the shaded arrow) may show up on the object system's past leg, as well as on the two future legs. Notice that the pattern of influence here is not simply the temporal inverse of that in the top left-hand diagram. In particular, the direction of influence on the agent's past leg does not change. This direction is provided by the agent's own temporal perspective, which is the same in both cases.

A complete reversal of the ordinary case would require the sort of pattern shown in the bottom left-hand diagram. In effect, this shows the normal case from the point of view of an agent whose orientation in time is the opposite of ours. (It is a picture of the normal pattern of causation in the opposite end of a Gold universe, as we might say.) Finally, the bottom right-hand diagram shows what the possibility of advanced action amounts to, for an agent whose orientation in time is the opposite of ours.

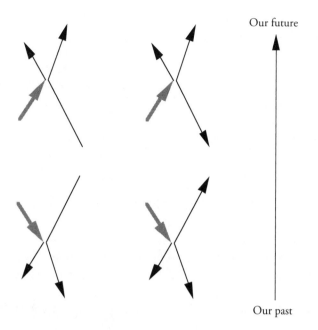

Figure 7.4. Four patterns of causal dependence.

So backward causation isn't simply forward causation in reverse. It is a more subtle possibility, thoroughly compatible with the discovery that the direction of causation contains a conventional or perspectival element: the possibility that the world is objectively patterned in such a way that from the ordinary asymmetric perspective, our influence seems to extend in the way shown in the top right-hand diagram, rather than simply in the way shown at the top left.

INVERTED FORKS AND DISTANT EFFECTS

Again, what would a symmetric world be like, on this view? For one thing, it would seem to give us the power to affect distant states of affairs indirectly, via an influence on the future factors on which those states depend. In terms of Figure 7.4, the suggestion is that our influence on one past leg allows us to affect the other. Changes to the local polarizer would influence photons which have yet to arrive, for example.

It might be thought that symmetry would require that we be able to do the analogous thing with the familiar future-directed forks. Shouldn't we be able to affect conditions on the distant leg of such a fork by influencing conditions on the near leg?

Well, yes and no. Yes in theory, in the sense that if we could influence conditions on the near leg *between ourselves and the vertex of the fork,* the fork structure should "impose" a corresponding influence on the distant leg. But no in practice, because we can't influence conditions on that part of the near leg. Taking the theoretical point first, we can perhaps imagine a creature who could exert influence on the past in the required way. For example we might imagine that God, in Her infinite wisdom, chose to create the world by first fixing the Earth and its inhabitants circa 1996, and then picking earlier, later, and more distant bits from among those compatible with that initial region of space and time. Suppose that on Earth in 1996 we "remember" that in 1976 we despatched a message from the solar system by rocket ship, but that in 1986 our records of the content of that important message were lost. We hope we said the right thing to our galactic neighbors, but we have no way of knowing. Couldn't God take pity on us, and as She pencils in the past, pick a history in which the pre-1986 records show that the message is what we would now like it to be? In setting the local conditions circa 1986, She determines the message itself in the distant spacecraft, for She knows that the spacecraft will "rendezvous" with Earth in 1976, and that when it does so the message it carries will match that shown in the Earthly records between 1976 and 1986. In choosing the state of those records She chooses the message.

The point of this little story is to show that from a purely logical point of view the two temporal directions are indeed treated equally by the advanced action proposal. If we could affect our past, perfectly ordinary correlations in future-directed forks would allow indirect influence on events at a space-like distance—exactly the sort of influence that past-directed forks give us, according to the advanced action view. If we could affect the settings of polarizers in the past, for example, we could affect that states of photon which have already passed through those polarizers. The *practical* distinction arises from our inability in practice to affect our ("direct") past, not from any temporal asymmetry in the correlational structure of the world. It arises from our own asymmetric orientation on paths between forks, in other words, not some from some global asymmetry in the pattern of forks themselves.

SUMMARY: SAVING THE BABY

This chapter began with the objection that the conventionalist approach to the asymmetry of dependence is both too weak, in failing to account for an objective asymmetry in the world, and too strong, in ruling out backward causation by fiat. To the first part of this objection I replied that conventionalism is as strong as it needs to be. It explains why the relevant asymmetry seems as objective as it does from our point of view, and this is good enough. To the second part of the objection I replied that conventionalism can indeed make sense of backward dependence, so long as it allows the crucial conventions to be appropriately sensitive to the ways in which we actually engage with the world. (Ordinary usage doesn't often encounter the issues concerned, however, and tends to be ambiguous at the relevant points.)

The latter reply has turned out to deliver a surprising and rather ironic bonus. In effect, it led us to this question: *What kind of world could coherently admit a convention according to which the past might be counterfactually dependent on the future?* This is a question about the world itself, about a thoroughly objective matter. Our interest in the subjective foundations of folk intuitions about causality and dependence thus led us to a live objective possibility—a possibility almost invisible to folk intuition, almost ignored by physics, and yet almost irresistible, I think, when weighed on scales whose judgment is not already compromised by temporal bias.

So surprising is this conclusion that we might well doubt that there could be a live objective issue of this kind. We might again try, as physicists often have, to insist on the "evident" temporal symmetry of physics, and hence to dismiss the present issue as being somehow an artifact of our subjective concepts of causality. But recall how we got to this point: we picked out the subjective strands in our thought about dependence, and then focused on the remainder. The great virtue of this approach is that it counters the temptation to put the anomaly of causal asymmetry into a basket labeled ANTHROPOCENTRISMS, which physics can safely ignore. For we have discovered that when we prune off the anthropocentrisms of the asymmetry of dependence, and put them away in that basket, an embarrassing anomaly remains in view.

In particular, the recognition of the subjectivity of causal asymmetry does not throw the baby out with the bath water, as I put it earlier, leaving no objective content to the advanced action view. The problem was simply that we didn't have a clear impression of the distinction between the water and the baby—that is, between our subjectively grounded intuitions concerning causal asymmetry and an issue concerning patterns of correlation in the world. Once the distinction has been drawn, however, it is possible with a little care to discard the dirty water and keep the baby.

The baby is a proposition concerning the correlational structure of the microworld, and we have seen that there are two very different forms this structure might take. There is an objective distinction between worlds which look as if they contain unidirectional causation and worlds which look as if they contain bidirectional causation (where "look as if" is to be filled out in terms of the conventionalist's account of how the temporally asymmetric nature of agents comes to be reflected in their concept of causation). With the benefit of hindsight we can see that we have never had good reason to exclude the structure that permits interpretation in bidirectional terms—on the contrary, if anything, given the symmetry argument. In order to be able to see this, however, it was essential to appreciate the subjective character of the interpretation.[16]

Our main project in the next part of the book will be to investigate the relevance of these conclusions for current issues in the interpretation of quantum mechanics. I want to show that advanced action has some striking advantages in quantum theory, of the kind foreshadowed in chapter 5. These advantages provide powerful new arguments in favor of advanced action, I think, but should not be allowed to overshadow the more general argument from symmetry. The symmetry point on its own provides a very strong reason for taking advanced action seriously; indeed, in my view, for regarding it as the proper default option for reality, from which departures need to be approved. With a certain amount of Archimedean gall, then, we might say that advanced action is just what we should have expected in the world, and that quantum mechanics simply shows us how God managed to pull it off!

8

←——————→

Einstein's Issue: The Puzzle of Contemporary Quantum Theory

E INSTEIN'S role in quantum mechanics provides some of the great ironies of twentieth-century physics. In the early years of the century, Einstein helped to lay the foundations for quantum theory. His work on the photo-electric effect, in which incoming light causes electrons to be released from a metallic surface, helped to establish that electromagnetic radiation has a particle-like aspect. A generation later, however, when quantum mechanics arose from these foundations, Einstein was to reject the new theory, or at least what became the standard view of its significance. Einstein—the revolutionary of those early years—became seen as the great reactionary, locked in a bitter dispute with Niels Bohr (1885–1962) about the meaning and the adequacy of quantum mechanics. Bohr carried the popular vote, but Einstein was never convinced, and the disagreement lasted for another generation, until Einstein's death in 1955.

Almost two generations later, however, the puzzles of quantum theory remain unresolved. The theory is without equal as a practical tool in physics, but its real significance remains profoundly unclear. For practical purposes this usually doesn't matter, and most working physicists are able to ignore the question as to what quantum theory is actually telling us about the world. When physicists and philosophers do address this issue, however, confusion reigns. There are no clear answers, and profound problems to be faced by all the evident contenders. As a result, the debate has become a kind of auction of lesser evils. In order to stay in the game, the competing players seek to make light of their own disabilities, and of course to interpret the rules in the most favorable possible fashion. Different players thus have different conceptions of the nature of the game.

What is an impartial spectator to make of this? It turns out that the best place to start is with the issue that Einstein himself saw as pivotal: *Does quantum mechanics provide a complete description of the physical world, or does it leave something out?* This is the main issue that divided Einstein and Bohr. Einstein always believed that quantum mechanics is not a complete description of the physical world, while Bohr and his Copenhagen allies disagreed. Two generations later, this issue—Einstein's issue—still allows a spectator to divide the players in quantum mechanics into two broad camps, and to identify the main problems that each camp must face, if it is to give us a satisfactory interpretation of quantum mechanics.

I am not an impartial spectator, of course. As I have said, I want to argue that much of the current perplexity about quantum mechanics stems from deeply ingrained but groundless intuitions about the temporal asymmetry of the microworld. I want to suggest that when these intuitions are challenged, it turns out that quantum mechanics is "the theory we had to have." At least in some of its most general and most puzzling aspects, in other words, it is the kind of theory of the microworld we would have expected, if we had properly explored these issues of temporal symmetry in advance.

In order to appreciate the advantages of advanced action in quantum mechanics, however, it is important to have a sense of the problems facing more conventional approaches. In order to appreciate its appeal, one needs a sense of the conceptual horrors it enables us to leave behind. In this chapter, then, I want to take on the role of an impartial spectator to the conventional debate—that is, the debate that ignores the possibility of advanced action. By focusing on Einstein's issue of the completeness of quantum mechanics, I want to mark out the main positions in the current spectrum of views about the significance of quantum theory, and especially to identify the characteristic difficulties of each position.

Impartiality has its limits, however, and the conclusion of this chapter is somewhat partial and unorthodox, even in the terms of the conventional debate. For one thing, Einstein's view that quantum mechanics is incomplete is often said to have been discredited by later developments. I think that this is a mistake, which results from a kind of dialectical double standard (though not, this time, the temporal double standard). Einstein's opponents apply an important constraint to his program; called "locality," this constraint amounts to a prohibition on action at a distance. However, while locality is a constraint that Einstein himself would like to have imposed— indeed, it receives much of its support from Einstein's own theory of special relativity—it is also one that almost all opposing views are unable to satisfy. Thus although a modified Einsteinian view without this constraint would have seemed unattractive from Einstein's own perspective, it cannot be excluded by proponents of opposing views.

Einstein's view will thus do rather better in the survey in this chapter than popular opinion might lead one to expect. All the same, I think that this defense of Einstein's intuitions about quantum mechanics is overshadowed by the possibility of an advanced action interpretation. As I want to argue in the next chapter, advanced action seems to vindicate Einstein's view *without sacrificing locality.* With the lessons of temporal symmetry properly considered, then, a very strong case can be made for the conclusion that Einstein was right: quantum mechanics is incomplete, though in a way in which Einstein himself seems never to have envisaged.

First, then, to the conventional debate. I should emphasize that the following sketch of contemporary views about the meaning of quantum mechanics is a *very* broad view. It ignores all but the prominent landmarks in the vast landscape that the contemporary debate about these problems has become. Even more than usual, I have had to ignore many details, in order to present the big picture in a reasonably accessible way.

To change the metaphor, I am going to ignore all but the most prominent branches of the contemporary debate about the meaning of quantum theory. If we want to step back far enough to see the major branches, we are bound to lose sight of the twigs. But there is also another motivation: eventually I want to call attention to some issues which lie right at the base of the tree. Roughly speaking, I want to suggest that the main puzzles of quantum mechanics arise because the theory has always been grafted onto diseased classical rootstock. The task is to prune away those diseased roots, and the distorted top growth which stems from them, in order that the theory might develop into its proper form on its own natural roots.

THE QUANTUM VIEW: BASIC ELEMENTS

Quantum mechanics characterizes a physical system by means of a so-called *state function* (or *wave function*), often symbolized by ψ. The empirical significance of the state function lies in the predictions it yields concerning observations, or measurements. Generally, the state function does not provide a unique prediction concerning the result of a possible measurement on the system in question. Instead, it assigns probabilities to a range of possible outcomes. This is the source of the claim that quantum mechanics is indeterministic. And one route—not the only one, as we shall see—to the issue of the completeness of quantum mechanics is provided by the question: Is this indeterminism fundamental, or is it simply a consequence of the fact that ψ doesn't tell us all there is to know about the system in question? Anyone who dislikes indeterminism is likely to be attracted to the latter possibility.

Quantum mechanics tells us how the state function of a system changes over time. Perhaps surprisingly, given its indeterministic reputation, the change it describes is mainly deterministic. The theory tells us that so long a system is left to its own devices, its state function varies continuously and deterministically, in accordance with a mathematical rule known as Schrödinger's Equation. The only exception to this principle concerns what happens at measurement. At this stage, at least according to the orthodox interpretation of the theory, the state function undergoes a sudden, discontinuous, and generally indeterministic change, the nature of which depends in part on the nature of the measurement in question. If we measure the position of a particle, for example, its new state reflects the fact that what was measured was position and not some other property. Indeed, the new state reflects the measurement result we actually obtained, in the sense that repeating the measurement will give the same result. The new state is thus "position-definite," meaning that it predicts with certainty the result of a new position measurement. If the state we started with was not already position-definite, however, then it is an indeterministic matter which position-definite state we end up with, after the initial position measurement.

In sum, then, quantum theory describes two "modes of evolution" for the state function—two ways in which the state function changes over time:

• When no measurement is being made on a system, its state function evolves smoothly and deterministically over time, in accordance with Schrödinger's Equation.

• When a measurement is made on a system its state function "collapses" discontinuously and indeterministically, yielding a new state function of a kind that depends on the nature of the measurement concerned.

Much debate about the interpretation of quantum mechanics has turned on the significance of these two modes of evolution. Physicists inclined to the view that quantum mechanics provides a complete description of reality have thought of them something like this: before we measure, say, the position of a particle, its position is normally *objectively* indeterminate. It is not simply that the particle has a position which we don't yet know—if that were so the state function wouldn't be telling us the complete story—but that it has no determinate position at all. When we perform a position measurement on the particle, however, its state function "collapses" in such a way as to ensure that its position becomes determinate—it acquires a definite position, which it didn't have before.

A TOM SPLIT IN THOUGHT EXPERIMENT!

The kind of view just described quickly became popular in the early days of quantum mechanics. It came to be called the *Copenhagen Interpretation,* in reference to the school led by Niels Bohr, who was the view's most prominent proponent. There were dissenting voices, however. One of the most famous of these was the voice that brought us what is perhaps the best-known thought experiment in the history of physics. Everyone knows the story: a cat is locked in a box with a deadly device, set to be triggered by the kind of microscopic event for which quantum mechanics yields indeterministic predictions. If a radioactive atom decays, for example, the cat will die. As Edwin Schrödinger (1887–1961) himself says,

> The typical feature of these cases is that an indeterminacy is transferred from the atomic to the crude macroscopic level, which can then be decided by direct observation. This prevents us from accepting a "blurred model" so naively as a picture of reality. By itself reality is not at all unclear or contradictory. There is a difference between a blurred or poorly focussed photograph and a picture of clouds or fog patches.[1]

Thus Schrödinger's thought is that *if* we say that before the box is opened and a measurement is made, the condition of the radioactive atom (whether or not it has decayed) is an objectively indeterminate matter, *then* we shall have to say the same about the condition of the cat—in other words, we shall have to say that the cat too is not determinately alive or dead until the box is opened. Schrödinger thinks that this latter conclusion is absurd, and hence that the view that quantum mechanics provides a complete description must itself be false. There must be determinate values in unmeasured systems, in other words, and quantum mechanics must be a blurry picture of a sharp reality, rather than the other way around.

The usual response to this argument is to deny that the state of the cat need be indeterminate before we open the box, by arguing that in effect a measurement is made by the cat itself, or by the apparatus which connects the atomic process to the lethal instrument. So the state function collapses well before the human experimenter opens the box, and never contains a cat in an indeterminate condition, neither alive nor dead. This response merely postpones the problem, however. A nice way to make the point is to consider a range of variants of the original experiment, in which the cat is replaced by other things. In one series of experiments we replace the cat with successively more complex and intelligent systems, culminating perhaps in a human observer. In another series we replace the cat with progressively simpler and "more stupid" recording devices, until the decay of the radioactive atom is affecting only a few other atoms.

We thus generate an array of experiments, in which the gradations from one to the next can be as fine as we like. In which of these experiments does a measurement occur—does the state function of the radioactive atom collapse—before the box is opened? At the simple end the entire apparatus in the box seems to count as an isolated system if anything ever does, and so we would expect its evolution to be governed by the continuous mode. At the complex end—indeed, already at the cat stage—we want to say that collapse does take place in the closed box. But where is there a place at which we can draw the line, at which we can plausibly argue that there is suddenly enough of a difference to flip the combined system in the box from one mode of behavior to the other?

It is important to this argument that we can't finesse the problem just described by saying that there are two *compatible* stances in all versions of the experiment. There are indeed two possible stances, namely the outsider's stance, from which the entire contents of the box appear isolated until the box is opened; and the stance of the recording instrument within the box, which "takes itself" to be making measurements on the quantum system concerned. The problem arises because, in virtue of the two modes of evolution described above, the two stances yield different (and *incompatible*) mathematical descriptions of the evolution of the combined system in the box. Most important, they differ on the issue as to when the discontinuous change takes place.

The difficulty I have just described is called the *measurement problem.* Essentially, it is the issue as to what counts as a measurement for the purposes of quantum mechanics. What makes the problem so intractable is that to the extent that quantum mechanics claims to be a universal theory, applicable to all physical systems, we should expect it to be applicable inter alia to those physical systems which may be used as measuring devices. Accordingly, we should expect the interaction between such a system and an object system on which a measurement is performed to be itself describable in quantum mechanical terms. As long as this interaction is not itself the object of a measurement, however, the theory appears to dictate that it should be governed by the continuous deterministic mode of evolution described by Schrödinger's Equation—and yet this seems flatly incompatible with the requirement that a discontinuous indeterministic change take place at the time of measurement.

I think it is fair to describe the measurement problem as an unsolved one, despite heroic efforts. There are a number of proposed solutions, each with an enthusiastic band of supporters, but no general consensus even on the likely form of a solution. I shall not attempt to do justice to these various proposals here. The main point I want to emphasize is that the measurement problem itself is entirely a consequence of the view that quantum mechanics provides a complete description of physical reality. This is easily illustrated in terms of Schrödinger's own metaphor. If the quantum mechanical state function is merely a fuzzy picture of a sharp reality, then the change that "takes place" when a measurement is made need reflect nothing more than a new viewpoint, the acquisition of a new picture showing some details that were not apparent in the old. In other words, it corresponds to a change in our information about the world, rather than a change in the world itself. The idea that a change in the picture requires a change in the world is simply a consequence of the assumption that the picture shows everything. Abandon this assumption, as Schrödinger himself recommends, and his cat becomes a toothless tiger. Just as there can be different incomplete photographs of the same thing, so there can properly be different incomplete descriptions of the contents of the box. Different perspectives yield different descriptions.

The view that Schrödinger is defending is sometimes called the *ignorance interpretation* of quantum mechanics, for it suggests that the so-called collapse of the wave function simply corresponds to a change in our degree of ignorance about the true state of the world, and not to a real physical change in the world. It is also referred to as the *hidden variable view,* since it holds that there are further facts—"hidden variables"—in the world, which are not given to us in the quantum mechanical description.[2]

Of course, defenders of the complete description view—Schrödinger's opponents—object at this point that there are profound objections to his own ignorance interpretation. This is quite true, and we shall come to these objections in a moment. But we mustn't make the mistake of thinking that our opponent's burden lightens our own. (It gives us an incentive to try to cope with our own burden, but that is quite a different matter.)

Before we turn to the problems of the ignorance interpretation, let's examine some further problems for the rival view. Schrödinger is not the most famous opponent of the complete description interpretation. The paper from which I quoted above was a response to (and appeared in the same year as) a famous 1935 paper by Einstein and two of his Princeton colleagues, Boris Podolsky and Nathan Rosen. In this paper—called "Can Quantum-Mechanical Description of Physical Reality be Considered Complete?"—Einstein presents his strongest attack on the complete description interpretation. And it is Einstein, not Schrödinger, who is remembered as the great opponent of the doctrine that quantum mechanics is complete.

THE EPR ARGUMENT

The EPR (Einstein-Podolsky-Rosen) paper introduces a class of experiments which turn out to involve some of the strangest consequences of quantum mechanics. Now known collectively as EPR experiments, the crucial feature of these cases is that they involve a pair of particles, or physical systems, which interact and then move apart. Providing the interaction is set up in the right way, quantum theory shows that the results of measurements on one particle enable us to predict the results of corresponding measurements on the other particle. For example we might predict the result of measuring the position of particle 1 by measuring the position of particle 2, or predict the result of measuring the momentum of particle 1 by measuring the momentum of particle 2.

This was the feature of these cases that interested Einstein, Podolsky, and Rosen. Einstein was seeking to undermine what was already becoming the orthodox interpretation of the fact that quantum theory appears to show that it is impossible to determine accurately and simultaneously both the position and the momentum of a physical system. As I noted above, the orthodoxy—the Copenhagen view—was that quantum systems do not have classical properties such as position and momentum, except when an appropriate measurement is made. Einstein wanted to argue that the restriction on measurement was merely epistemological, however. That is, he thought that it was merely a restriction on our knowledge of the physical world, rather than on reality itself.

In philosophical terms, Einstein was a realist: he believed that the world exists independently of minds and observations. Hence he bitterly disliked Bohr's view that the nature of reality depends on what humans choose to observe. He believed that the features of quantum mechanics that Bohr and others took as evidence of deep entanglement between observation and reality were really a reflection of the fact that the theory gives only a partial description of reality. As he saw, then, the crucial question is therefore whether the quantum mechanical description of reality can be considered to be complete. Does it say all there is to be said about a physical system, or are there further facts about the physical world not captured by quantum mechanics?

The two-particle systems seemed to provide the decisive argument that Einstein was looking for. With Podolsky and Rosen, he argued that the existence of such systems showed that quantum theory must indeed be incomplete. For if we can predict either the measured position or the measured momentum of a particle without interfering with it in any way, then it must have some property responsible for the results of those (possible) measurements. If we measure the position of particle 2 we infer that particle 1 has a definite position. If we measure the momentum of particle 2 we infer that particle 1 has a definite momentum. But since in neither case do we do anything to particle 1 itself, it must have a definite position and momentum—regardless of what we *actually* do to particle 2. In other words, particle 1 must have properties not described by quantum mechanics. Quantum mechanics must be incomplete.

The EPR argument failed by and large to sway supporters of the Copenhagen Interpretation, but this is perhaps due more to the obscurity of the Copenhagen response than to any compelling counterargument it brought to light. With the benefit of hindsight we would probably now say that Einstein was right, had John Bell not unearthed a remarkable sting in the tail of the EPR experiment, some thirty years later. We shall come to Bell's argument later on. Briefly, however, Einstein, Podolsky, and Rosen had assumed that what we choose to measure on particle 2 could not have an effect on the distant particle 1. For example, measuring the position of particle 2 could not somehow "bring it about" that particle 1 had a definite position. In other words, the EPR argument for the incompleteness of quantum mechanics assumes that physical effects are *local*—that there is no action at a distance. The sting revealed by Bell's work is that other features of the quantum mechanical description of certain EPR cases seem to show that any more complete theory would have to be nonlocal. It would have to reject the very assumption on which the EPR argument depends. Einstein's allies thus find themselves in an unfortunate dilemma. To make a hidden variable theory work—to make it consistent with quantum mechanics—they have to abandon the assumption that enabled them to argue from the possibility of the EPR experiment to the conclusion that there must be such a theory.

However, it is very important to appreciate that this does not show that Einstein was wrong. It simply saves his opponents from what would otherwise be a very serious objection. A common misconception is that Bell's argument excludes hidden variable theories *tout court*. It does not. Even leaving aside the loophole provided by advanced action, Bell's result counts only against *local* hidden variable views, leaving open the possibility of nonlocal hidden variable theory—while nonlocality itself can hardly be held to be a decisive failing in hidden variable views if other views need it as well, as Bell's result suggests that they do.[3]

We'll return to the issue of locality in a moment. Before we do so, however, we should note that even if we leave to one side the original EPR argument, there is a more intuitive point in favor of hidden variables to be drawn from the kind of examples on which the argument relies. For suppose it is true that the particles concerned have no definite positions and momenta until and unless they are subject to an appropriate measurement—at which stage a definite value of the appropriate quantity "materializes" from an indeterminate fog of possibilities. If there is nothing "there" in reality except the fog, why is it that what materializes at one side always stands in the same relation to what materializes at the other? Why is there such a constraint on what would appear to be distinct indeterministic processes? There seems to be a puzzle here which could only be resolved by adding something to the formalism of quantum mechanics—in other words, by conceding that quantum mechanics does not provide a complete description.

A defender of the complete description view is likely to reply that the processes are not independent, as this argument seems to assume. It will be said that because the particles have interacted, they are governed by a single combined state function, of which it is a consequence that the materialization of definite values obeys this constraint. But doesn't this simply restate the problem? The state function tells us *that* possible results of measurements are correlated, but not *why* this should be so. It gives us a *description* of the correlation, but not an *explanation*. The complete description view seems to amount to the assertion that with the state function we reach bedrock: there is no further explanation possible of *why* things are like this. The appropriate response seems to be to say that while of course we cannot rule out the possibility that we might reach bedrock at a given stage in physics, it is a prima facie disadvantage of a theory that it requires us to concede that we have actually done so. (It is something like the inability to move one's pieces in a game in which this does not result in stalemate. So long as one's opponent's pieces are free, it is foolish to paint oneself into a corner in this way.)

EPR AND SPECIAL RELATIVITY: THE COST OF NONLOCALITY

Returning now to the issue of locality, we find that when we take into account Einstein's first and perhaps most famous contribution to physics, a somewhat sharper objection to the complete description view may be extracted from the EPR cases. The view is committed to the claim that measurements on one particle of an EPR pair affect the state function of the combined system, no matter how far apart the particles happen to be. In the right circumstances a measurement *here* produces a real change in the world *there*, no matter how far away "there" is. But when does the influence arrive? The correlations to which the EPR argument appeals hold even when the relevant measurements lie at a spacelike distance from one another, in the sense of special relativity— that is, when they are inaccessible to each other by means of a signal traveling at or below the speed of light. If we measure the position of particle 1 we can predict the result of a position measurement on particle 2, even though the latter measurement takes place at a spacelike distance. If the state function relevant to the prediction of the results of measurements on particle 2 had not changed at this stage, we would know something about particle 2 which was not given to us by a knowledge of the state function. In other words, the state function would not be a complete description. So it seems that the "effect"—the collapse of the state function—cannot be constrained by the usual lessons of special relativity.

This concession leads to new problems, however. Unless there is a privileged reference frame in terms of which all such collapses are instantaneous, we are likely to run into consistency problems. The order of events *at a given location* will appear different from different points of view. Suppose we measure the position of particle 1 and the momentum of particle 2, the two measurement events being spacelike related to one another. An observer whose view is that the position measurement takes place before the momentum measurement will say that the system moves from its original state ψ to one in which the two particles have position-definite states, and that particle 2 later moves to a momentum-definite state. (The first measurement destroys the correlation between the momenta of the two parts of the system, so that the second measurement does not result in a momentum-definite state for particle 1. After the measurement on particle 2, particle 1 appears from this perspective to remain in a position-definite state.) An observer whose view is that the momentum measurement occurs first will "see" a different course of events: from her point of view the original combined state ψ will give way to momentum-definite states of the two particles. Later, the position measurement on particle 1 gives that particle a position-definite state, while particle 2 remains in a momentum-definite state. The two observers will disagree, for example, as to whether there was ever a stage at which a position measurement on particle 2 would have yielded a definite result. Again, this seems incompatible with the view that the state function encapsulates the complete truth about a physical system. The two perspectives yield different accounts of the "complete truth" concerned. (Note that this isn't like special relativity itself, where the theory does provide a picture of the objective reality which underlies the frame-dependent properties of mass, length, and time.)[4]

In sum, the EPR argument continues to present grave problems for a complete description view of quantum mechanics, despite the apparent failure of the argument in its original form. The original argument assumed locality, and Bell's Theorem is generally taken to establish that this assumption is untenable. But nonlocality is not a problem for hidden variable theories alone. It is difficult to see how it can be accommodated by a complete description view, without rejecting one of the fundamental principles of special relativity, that there is no such thing as absolute simultaneity. This problem aside, there is a disturbing residue in the complete description view's account—or lack of an account—of what goes on in EPR cases. A view that holds that measurement results simply materialize by chance from an indeterminate fog of possibilities seems particularly ill-placed to explain why distant parts of the fog should display such precise correlations.

Let me emphasize once again that in claiming that the EPR cases are problematic for complete description views, I am not suggesting that they are unproblematic for the rival hidden variable views. The sins of one's enemies do not atone for one's own. This is a lesson that both sides in the conventional debate about quantum mechanics need to take to heart, however.

We shall come to the difficulties facing hidden variable views in a moment. I want first to mention one more difficulty for the complete description interpretation, which is one I have already touched on earlier in the book.

THE TEMPORAL ASYMMETRY OBJECTION

According to the orthodox version of the complete description account, the state of a quantum system in the period between one measurement and the next reflects the nature of the previous measurement. When we measure the position of a particle, for example, its state then evolves in accordance with Schrödinger's Equation, from a starting point which reflects the fact that the particle has been found to have a well-localized position. Its state does not evolve in a way which reflects what is to happen to it in the future, however. If the next measurement involves a determination of its momentum, for example, this is not reflected in its state between the two measurements.

The asymmetry here is easily displayed by the device we used in earlier chapters, of looking at the case from the reverse of the ordinary temporal perspective. From this viewpoint we see a particle evolving *toward* a state which is associated with a particular kind of interaction in which it is to be involved in the future (the interaction that from the ordinary viewpoint we describe as a position measurement). The fact that this now looks very odd is a good indication that there is a temporal asymmetry in play here, which we are ordinarily inclined to overlook. (As before, we are taking advantage of the fact that asymmetries too familiar to notice are thrown into stark relief when we look at things in reverse.) We ordinarily take for granted that it is natural for the state to depend on the past in a way in which is does not depend on the future. But with what justification?[5]

The discussion of the origins of temporal asymmetry earlier in the book suggests that there are two ways in which this asymmetry might turn out to be relatively unproblematic. The first would be for it to turn out to be of the same origin as the thermodynamic asymmetry, but this does not seem likely. After all, the complete description view requires that it characterizes the intrinsic behavior of individual systems, rather than being a statistical feature of large collections of systems. Could it perhaps be an anthropocentric asymmetry, of the kind we examined in chapters 6 and 7? This would be the second unproblematic alternative, but again it seems incompatible with the spirit of the complete description view. A good way to confirm this is to return to our simple photon case. In the orthodox quantum mechanical description the state of the photon between the polarizers is correlated with the orientation of the earlier polarizer but not that of the later polarizer.[6] If the state itself is an objective property of the system, this difference reflects an objective asymmetry in the structure of reality. (As we saw in chapter 7, this is important respect in which photons—at least as commonly conceptualized—differ from Newtonian billiard balls.)

Again, it is important to appreciate that this difficulty is a consequence of the view that the state function is a complete description. According to the ignorance interpretation, the asymmetry might simply reflect the fact that we know more about the history of the system in question than we do about its future. Indeed, we might see the state function as a kind of summation of what could be known about a system in virtue of a complete knowledge of its past interactions with other things. In this case, the asymmetry would be no different from that of many familiar cases in which we describe something in an incomplete way.

Consider a medical investigation, for example. A patient consults his doctor, who takes a history and begins a series of tests. At each stage the doctor is able to describe her patient in terms which reflect the results of investigations already completed, but not those which are yet to come. Early in the process, for example, the doctor might describe her patient as a thirty-five-year-old male, of average weight, with high blood pressure and a recent history of stomach pain. She won't yet describe him as someone with a fish bone wedged in his esophagus, because the X-ray which reveals this has yet to be performed. But this temporal asymmetry in the description—the fact that at each stage it depends on observations in the past but not those in the future—reflects an asymmetry in our process of knowledge acquisition, not any objective asymmetry in the subject matter. The patient had the fish bone all along, and didn't change when the doctor discovered it. The change occurred in the doctor's state of information, not in the world described.

In the case of quantum mechanics, then, the temporal asymmetry involved in the evolution of the state function as a result of measurement—the fact that the state depends on the previous measurement but not the next measurement—is not problematic if the state function is thought of as an incomplete description. It is only a difficulty for a complete description interpretation of the state function, for on this view it represents an objective temporal asymmetry in the world.

This objection to complete description interpretations has received rather scant attention in the vast literature about the foundations of quantum mechanics. This is hardly surprising, however, if we bear in mind that it is simply an example of the intuitively appealing objective asymmetries we have been looking at in the last three chapters. In effect, this asymmetry is simply a concrete manifestation of μInnocence, or the asymmetry we identified in the photon case. One of the lessons of those cases was that the asymmetries concerned are so familiar, so intuitive, and so easily confused with what turn out to be the subjectively grounded asymmetries of causation, that it is very difficult to see that they do conflict with plausible principles of temporal symmetry.

In the quantum mechanics case, another source of confusion is that the best-known attempt to address the problem of the apparent time asymmetry of quantum measurement does not actually touch on the main difficulty. In a paper published in 1964, Yakir Aharonov, Peter Bergmann, and Joel Lebowitz set out to "examine the assertion that the 'reduction of the wave packet,' implicit in the quantum theory of measurement introduces into the foundations of quantum physics a time-asymmetric element, which in turn leads to irreversibility."[7] They go on to argue that this asymmetry "is actually related to the manner in which statistical ensembles are constructed." In seeing this as a route to a solution, however, they are taking for granted that in applying a quantum state to a system, we simply describe it as belonging to a certain ensemble or class of similar systems. And this is an *incomplete* description view of quantum theory: it is just like the view that in describing someone as a thirty-five-year-old male with stomach pain, we locate that person in the class of all people who meet this description. Another measurement may enable us to locate the quantum system or the person in a smaller class, on this view, *with no change in the system or person itself.*

This is a view of quantum mechanics that Einstein would have found congenial, and Aharonov, Bergmann, and Lebowitz are quite correct to argue that if quantum mechanics is understood in this way, the time asymmetry of the measurement process is not problematic. What they fail to note, however, is that their argument does nothing to address the problem for those who disagree with Einstein—those who think that the state function is a complete description, so that the change that takes place on measurement is a real change in the world, rather than merely a change in our knowledge of the world. Had this been made explicit, later writers, not themselves in favor of Einstein's view, would have been less likely to make the mistake of thinking that Aharonov, Bergmann, and Lebowitz have solved the problem of the time asymmetry of measurement, as it effects the complete description interpretation.

In summary, then, we have identified three main problems for complete description interpretations of quantum mechanics: the measurement problem, the issues arising from the EPR experiment, and the temporal asymmetry problem. The first and last of these appear to be peculiar to the complete description approach, in the sense that it is relatively clear how they are avoided by an ignorance interpretation. The problem of nonlocality is different, in that as we are about to see, nonlocality seems equally to be a problem for hidden variable approaches. It is fair to say that the measurement problem is the most serious of these difficulties. Indeed, I think it is fair to say that it is so serious that the complete description approach would simply not be regarded as a serious contender, if there seemed to be a viable alternative.

In practice, the complete description view tends to be seen not merely as a serious contender, but as the only plausible account of what quantum mechanics is telling us about the world. The rival view, favored by Einstein, is often thought to be completely discredited. Why is this so? Given that the complete description view is itself so problematic, what terrible sins has the ignorance interpretation committed, to be rejected so thoroughly?

These days, there is a widespread consensus that the strongest argument against Einstein's approach is that provided by the work of John Bell, which builds on the foundation provided by the original EPR argument. Bell's work dates from the mid-1960s, however, by which time the complete description view had long been the orthodoxy. Before we turn to Bell's Theorem, then, we should ask what factors were influential in the early decades on quantum theory. Why did Bohr and his contemporaries reject Einstein's interpretation of quantum mechanics? And what weight do these factors still carry, independently of Bell's result?

THE CONSEQUENCES OF SUPERPOSITION

The early arguments against hidden variables turn on a feature of quantum mechanics known as *the superposition principle.* This says that if ψ_1 and ψ_2 are permissible wave functions for a quantum mechanical system, then so is any linear combination of ψ_1 and ψ_2—that is, any wave function of the form $a\psi_1 + b\psi_2$, where a and b are arbitrary complex constants.

Some of the puzzling consequences of superposition are exemplified in the famous *two-slit experiment,* which dates from the early days of quantum theory. In this experiment a beam of particles—photons or electrons, for example—is directed from a source toward a screen containing two slits, through which the particles may pass. Particles passing the screen are detected on a photographic plate, the sum of the impacts yielding a frequency distribution over the positions in the plane of the plate. The experiment may be run with one or both slits open.

Quantum mechanics predicts that the frequency distribution which results from the experiment with both slits open is *not* the sum of the distributions resulting from running it with each of the two slits open individually. This appears to imply that the particles concerned cannot be considered to have definite trajectories—precise positions, even when not observed. For if each particle did have such a trajectory, it could only pass through one slit or the other, and apparently its behavior on doing so should not depend on whether or not the other slit is open. So the frequency distribution for the two-slit case should resolve into two distributions: one due to those particles which in fact went through the upper slit, and one due to those which in fact went through the lower one. And each of these distributions should be equivalent to that obtained when only the slit concerned is open.

Early in the history of quantum mechanics, cases of this kind seem to have been largely responsible for fostering the view that quantum measurement does not so much reveal a preexisting reality as create one from an indeterminate fog of possibilities. As we have seen, however, there were dissenting opinions. The project of these dissenters was to try to show that the predictions of quantum mechanics could be reproduced by a model in which the state function was not a complete description—a model, in other words, in which there were further elements of reality whose values could account for the results of measurements on quantum mechanical systems.

The usual conception of this project involves two main ideas. The first is that the model should account for the result of any possible measurement which might be performed on a given system, in terms of the system's underlying state at the time in question. The second is that it should explain the probabilistic predictions of quantum mechanics, in terms of a probabilistic distribution of the underlying states.[8] However, a series of mathematical results, culminating in that of S. Kochen and E. P. Specker in 1967,[9] seemed to show that no such hidden variable model is possible in quantum mechanics, at least in general. In effect, Kochen and Specker show that there are simple quantum mechanical systems for which no such set of underlying states exists.

Something of the flavor of results of this kind is conveyed by an informal analogy which is often used to illustrate the consequences of superposition. The story asks us to imagine a shell game, in which we are allowed to turn over any two of three up-turned cups. Whenever we do so, we always find one black stone and one white stone. Can this result be explained in terms of some prior distribution of stones among the three cups? A little thought shows that it cannot be. The distribution would need to place one and only one stone under each cup, for we never find more or less than this. Each stone would have to be either black or white, for we never find any other color. So there would have to be at least two stones of the same color, which contradicts the fact that we always find stones of different colors.

A large literature has grown up seeking ways to evade these *no hidden variable theorems,* as the Kochen-Specker result and others of the same kind are called. One of the assumptions of Kochen and Specker's result is that in satisfying the requirement that the model account for the result of any possible measurement, the results predicted by the underlying state for a given measurement M should not depend on what other measurements are made in conjunction with M. (In terms of the shell game analogy, this amounts to the requirement that the apparent color of the stone we find under a given cup does not depend on what other cup we turn over at the same time.) This assumption turns out to be essential. For various types of quantum mechanical systems, hidden variable theories lacking this property have been devised. In these so-called *contextual* hidden variable theories, the result of a given measurement depends not only on the underlying state, but also on what class of measurements, in addition to the given one, are being performed on the system in question.

The best-known example of a contextual hidden variable theory is that of David Bohm. Bohm's theory was first developed in the early 1950s, but was largely ignored by mainstream opinion in quantum mechanics. As David Albert remarks in a recent introduction to Bohm's ideas, many physicists seem to have dismissed the theory on the grounds that the no hidden variable theorems had shown that what Bohm had achieved was actually impossible![10] Bohm had a few notable supporters, however, including John Bell himself, who reformulated the theory in the early 1980s.[11]

There is some irony in Bell's support for Bohm's views. In recent years, as the no hidden variable theorems have been better understood, the conventional opinion has been that the contextual path might indeed provide a viable escape route for hidden variable approaches, were it not for the fact that Bell's Theorem rules out even contextual approaches. As Bell's support for Bohm's theory makes clear, however, he himself did not subscribe to this interpretation of the significance of his own results.

Before we turn to Bell's Theorem, I want to draw attention to a more basic assumption of the no hidden variable theorems than that challenged by the contextualist approach. It is the assumption that the preexisting hidden states do not depend on the *future* measurement interactions in which the system in question is to be involved. In terms of our familiar example, this is simply the assumption that the true state of the photon does not depend on the setting of the future polarizer. We have seen that this assumption contains an objective residue, not canceled by the recognition of the anthro-pocentric ingredients in ordinary talk of dependence. We have also seen that this residue is highly problematic. It is difficult to see why we should assume that the world is asymmetric in this way.

Contextualist approaches challenge the no hidden variable theorems by multiplying the variables. In effect, they postulate a multiplicity of new hid-den variables, tailor-made for the various combinations of measurements that quantum mechanics allows. By multiplying aspects of reality in this way, they are able to account for what quantum mechanics predicts about what we see when we look at reality in various different ways. What remains intact, however, is the idea that observation simple "reads off" some aspect of a preexisting reality. The alternative strategy I want to recommend in chapter 9 is that we abandon this benign classical view of measurement, and with it the assumption that a single hidden state need reproduce the results of any possible measurement on the system in question. If the stones know in advance which cups are to be turned over, it is a trivial matter to have them arrange themselves so that one of each color appears.

BELL'S THEOREM

I noted earlier that Einstein's attack on the complete description interpre-tation is often thought to have fallen victim to a sting in its own tail. The original EPR argument turns on the predictions that quantum theory makes about the correlations between the results of the various possible measure-ments on certain two-particle systems. However, Einstein, Podolsky, and Rosen had failed to notice some further features of these two-particle cor-relations, features whose significance was first noticed by John Bell in the mid-1960s. Bell examined a variant of the original EPR case—a version which had originally been described by David Bohm. In this new case, Bell uncovered what has turned out to be one of the most puzzling discoveries in the history of physics.[12]

Fortunately for lay readers, we don't need to know the details of Bohm's case to appreciate the puzzling character of Bell's conclusions. So long as we are prepared to take on trust the predictions of quantum mechanics, the essential features of Bell's argument can be described in terms of informal and much more commonplace analogies. Very little mathematical thought is required to show that if analogous correlations were to arise in familiar regions of the world, they would strike us as very odd indeed. Accordingly, Bell's Theorem has become the subject of a number of edifying parables. The following tale is loosely adapted from those given in several lucid and entertaining papers by the physicist David Mermin.[13]

Ypiaria 1: the Twin Paradox

By modern standards the criminal code of Ypiaria[†] allowed its police force excessive powers of arrest and interrogation. Random detention and questioning were accepted weapons in the fight against serious crime. This is not to say that the police had an entirely free hand, however. On the contrary, there were strict constraints on the questions the police could address to anyone detained in this way. One question only could be asked, to be chosen at random from a list of three: (1) Are you a murderer? (2) Are you a thief? (3) Have you committed adultery? Detainees who answered "yes" to the chosen question were punished accordingly, while those who answered "no" were immediately released. (Lying seems to have been frowned on, but no doubt was not unknown.)

To ensure that these guidelines were strictly adhered to, records were required to be kept of every such interrogation. Some of these records have survived, and therein lies our present concern. The records came to be analyzed by the psychologist Alexander Graham Doppelgänger, known for his work on long distance communication. Doppelgänger realized that among the many millions of cases in the surviving records there were likely to be some in which the Ypiarian police had interrogated both members of a pair of twins. He was interested in whether in such cases any correlation could be observed between the answers given by each twin.

As we now know, Doppelgänger's interest was richly rewarded. He uncovered the two striking and seemingly incompatible correlations now known collectively as *Doppelgänger's Twin Paradox*. He found that

(8·1)   When each member of a pair of twins was asked the same question, both always gave the same answer;

and that

(8·2)   When each member of a pair of twins was asked a different question, they gave the same answer on close to 25 percent of such occasions.

It may not be immediately apparent that these results are in any way incompatible. But Doppelgänger reasoned as follows: 8·1 means that whatever it is that disposes Ypiarians to answer Y or N to each of the three possible questions 1, 2, and 3, it is a disposition that twins always have in common. For example, if YYN signifies the property of being disposed to answer Y to questions 1 and 2 and N to question 3, then correlation 8·1 implies that if one twin is YYN then so is his or her sibling. Similarly for the seven other

---

[†]Pronounced, of course, "E-P-aria."

possible such states: in all, for the eight possible permutations of two possible answers to three possible questions. (The possibilities are the two homogeneous states YYY and NNN, and the six inhomogeneous states YYN, YNY, NYY, YNN, NYN, and NNY.)

Turning now to 8·2, Doppelgänger saw that there were six ways to pose a different question to each of a pair of twins: the possibilities we may represent by 1:2, 2:1, 1:3, 3:1, 2:3, and 3:2. (1:3 signifies that the first twin is asked question 1 and the second twin question 3, for example.) How many of these possibilities would produce the same answer from both twins? Clearly it depends on the twins' shared dispositions. If both twins are YYN, for example, then 1:2 and 2:1 will produce the same response (in this case, Y) and the other four possibilities will produce different responses. So if YYN twins were questioned at random, we should expect the same response from each in about 33 percent of all cases. Similarly for YNY twins, for YNN twins, or for any of the other inhomogeneous states. And for homogeneous states, of course, all six possible question pairs produce the same result: YYY twins will always answer Y and NNN twins will always answer N.

Hence, Doppelgänger realized, we should expect a certain minimum correlation in these different question cases. We cannot tell how many pairs of Ypiarian twins were in each of the eight possible states, but we can say that whatever their distribution, confessions should correlate with confessions and denials with denials in at least 33 percent of the different question interrogations. For the figure should be 33 percent if all twins are in inhomogeneous states, and higher if some are in homogeneous states. And yet, as 8·2 describes, the records show a much lower figure.

Doppelgänger initially suspected that this difference might be a mere statistical fluctuation. As newly examined cases continued to confirm the same pattern, however, he realized that the chances of such a variation were infinitesimal. His next thought was therefore that the Ypiarian twins must generally have known what question the other was being asked, and determined their own answer partly on this basis. He saw that it would be easy to explain 8·2 if the nature of one's twin's question could influence one's own answer. Indeed, it would be easy to make a total anticorrelation in the different question cases be compatible with 8·1—with total correlation in the same question cases.

Doppelgänger investigated this possibility with some care. He found, however, that twins were always interrogated separately and in isolation. As required, their chosen questions were selected at random, and only after they had been separated from one another. There therefore seemed no way in which twins could conspire to produce the results described in 8·1 and 8·2. Moreover, there seemed a compelling physical reason to discount the view that the question asked of one twin might influence the answers given by another. This was that the separation of such interrogations was usually spacelike in the sense of special relativity; in other words, neither interrogation occurred in either the past or the future light cone of the other. (It is not that the Ypiarian police force was given to space travel, but that light traveled more slowly in those days. The speed of a modern carrier pigeon is the best current estimate.) Hence according to the principle of the relativity of simultaneity, there was no determinate sense in which one interrogation took place before the other. How then could it be a determinate matter whether interrogation 1 influenced interrogation 2, or vice versa?[14]

How are we to explain Doppelgänger's remarkable observations? Doppelgänger himself seems reluctantly to have favored the telepathic hypothesis—the view that despite the lack of any evident mechanism, and despite the seeming incompatibility with the conceptual framework of special relativity, Ypiarian twins were capable of being influenced instantaneously by their sibling's distant experiences. As we shall see, however, Doppelgänger was aware that there is a hypothesis that explains 8·1 and 8·2 without conflicting with special relativity. It is that the twins possess not telepathy but precognition, and thus know in advance what questions they are to be asked. But he seems to have felt that this interpretation would force us to the conclusion that the Ypiarian police interrogators were mere automatons, not genuinely free to choose what questions to ask their prisoners. Other commentators have dismissed the interpretation on different grounds, claiming that it would give rise to causal paradoxes.

In my view neither of these objections is well-founded. The relativity-friendly alternative that Doppelgänger rejects is certainly counterintuitive, but it is not absurd. Given the nature of the case, any workable explanation will be initially counterintuitive. What matters is whether that intuition withstands rigorous scrutiny, and of course how much gain we get for any remaining intuitive pain. Doppelgänger himself seems to have been aware of the gains that would flow from the interpretation in question (especially that it saves special relativity), but thought the pain too high. When we return to this story later on, I want to try to show that he was mistaken.

The point of the Ypiarian example lies, of course, in the fact that it exactly mirrors the puzzling behavior of certain two-particle quantum-mechanical systems. Doppelgänger's 8·1 is effectively the feature of EPR systems on which the original EPR argument relied. And 8·2 is the additional feature whose conflict with 8·1 was noted by Bell in 1965. The case mirrors Bohm's version of the EPR experiment. The pairs of twins correspond to pairs of spin-1/2 particles in the so-called singlet state. The act of asking a twin one of three specified questions corresponds to the measurement of the spin of such a particle in one of three equi-spaced directions perpendicular to the line of flight. The answers Y and N correspond on one side to the results "spin up" and "spin down" and on the other side to the reverse. Thus a case in which both twins give the same answer corresponds to one in which spin measurements give opposite results. Correlations 8·1 and 8·2 follow from the following consequence of quantum mechanics: when the orientation of two measuring devices differ by an angle α then the probability of spin measurements on each particle yielding opposite values is $\cos^2(\alpha/2)$. This probability is 1 when $\alpha = 0°$ (8·1) and $1/4$ when $\alpha = \pm120°$ (8·2).

These predictions have been confirmed in a series of increasingly sophisticated experiments.[15] Thus if you thought the proper response to the Ypiaria story was that it was unrealistic, you should think again. Not only is the perplexing behavior of Ypiarian twins theoretically and practically mirrored in quantum mechanics, but quantum mechanics actually tells us what sort of neurophysiology would make people behave like that. All we have to suppose is that the brains of Ypiarian twins contain the appropriate sort of correlated spin-1/2 particles (one particle in each twin), and that interrogation causes a spin determination, the result of which governs the answer given.

Bell's result is commonly regarded as providing a decisive objection to the incomplete description approach to quantum mechanics. Bell is seen as the man who proved Einstein wrong, in the words of John Gribbin's tribute to Bell, after Bell's death in October 1990.[16] As I noted earlier, however, this interpretation seems a little one-sided. Leaving aside the possibility of advanced action, Bell's Theorem establishes that in order to explain what quantum mechanics predicts about EPR cases, a hidden variable theory would need to be nonlocal. In particular, it would need to allow that the hidden variables on one side of an EPR apparatus could be influenced by the measurement made on the other. While this conflicts with one of the two crucial assumptions of the original EPR argument for the incompleteness of quantum mechanics, it does not show that quantum mechanics is complete. It merely shows that in order to reproduce the quantum mechanical predictions, a hidden variable theory would need to be nonlocal.

Moreover, since complete description views also seem to embody nonlocality, it can hardly be held up as a decisive objection to the hidden variable approach. The right conclusion seems to be that to the extent that nonlocality is a defect, both views are equally at fault; while to the extent that it is not a defect, Bell's Theorem does nothing to undermine the hidden variable approach.

If the case against hidden variables is any stronger than this, it must rest on the claim that nonlocality is more problematic for Einstein's hidden variable approach than for its Copenhagen opponents. It is difficult to see why this should be so, however. No doubt it is true that those on Einstein's side of the debate have been more troubled by nonlocality than their opponents, but I think that this is only because the Copenhagen view has tended to be protected by the comfortable cushion of obscurity which surrounds its commitments concerning the nature of reality.

EPR FOR TRIPLETS: THE GHZ ARGUMENT

There is an interesting new class of Bell-like results in quantum mechanics, known as the Greenberger-Horne-Zeilinger (GHZ) cases, which seem to achieve Bell's conclusions by nonstatistical means. Some writers have seen these results as strengthening the Bell-based argument against hidden variables. I want to look briefly at these cases. Apart from their intrinsic interest, it will be important to be able to show that they do not provide any new obstacle to the strategy I want to recommend in chapter 9. On the contrary, I think, they simply strengthen the case for advanced action.

The GHZ argument lies somewhere between the algebraic no hidden variable theorems and Bell's Theorem. It combines the combinatorial reasoning of the former with the multiparticle insights of the latter, and thus achieves what might be termed Bell without statistics. Fortunately, the central combinatorial argument is even more straightforward than the core of Bell's Theorem.[17] Even more fortunately, it turns out that here, too, the fabled Doppelgänger has been here first.

Ypiaria 2: the triplet tapes

Tired of twins, Doppelgänger moved on to triplets. To his surprise, he found that the Ypiarian criminal code embodied special exemptions for triplets, who were excused for adultery on grounds of diminished childhood responsibility. When subject to random interrogation, then, triplets were only asked one of the two questions (1) Are you a thief? and (2) Are you a murderer? In records of these interrogations, Doppelgänger found that on occasions on which all three members of a set of triplets were questioned in this way, their answers always conformed to the following pattern:

(8.3) When all three triplets were asked the first question, an odd number of them said "no."

(8.4) When two triplets were asked the second question and one the first, an even number (i.e., two or none) of them said "no."

As in the twins case, Doppelgänger asked himself whether it is possible to explain these results in terms of local hidden variables—that is, in terms of psychological factors predisposing each triplet to respond in a certain way to either of the possible questions, independently of the concurrent experiences of his or her fellow triplets. Reasoning as follows, he decided that this was impossible.

Suppose that there are such psychological factors. Given a particular set of triplets, let us write a_1 and b_1 for the factors responsible for the answer the first triplet would give to the first and second question, respectively, and similarly a_2, b_2, a_3, and b_3 for the corresponding factors in the second and third triplets. And let us think of these factors as having values +1 or –1, according to whether they dispose to a positive or negative answer. Then 8·3 implies that the product of a_1, a_2, and a_3 is –1 (since it contains an odd number of negative factors); and 8·4 implies that each of the products $a_1 b_2 b_3$, $b_1 a_2 b_3$, and $b_1 b_2 a_3$ has value +1 (since it contains an even number of negative factors). Taken together these results in turn imply that the combined product

$$(a_1 a_2 a_3)(a_1 b_2 b_3)(b_1 a_2 b_3)(b_1 b_2 a_3) = (-1)(+1)(+1)(+1) = -1.$$

This is impossible, however, since each individual factor occurs exactly twice on the left hand side, so that negative factors must cancel out. Hence local hidden variables cannot account for the observed results.

Our present interest in this case lies in the fact that the triplet correlations that Doppelgänger discovered in the Ypiarian case exactly match those predicted by the recent GHZ results in quantum mechanics. The behavior of Ypiarian triplets parallels that of sets of three spin-1/2 particles in the so-called GHZ state, when subject to combinations of spin measurements in one of two chosen directions perpendicular to their lines of flight. Particle 1 can thus have its spin measured in direction a_1 or direction b_1, particle 2 in direction a_2 or b_2, and particle 3 in direction a_3 or b_3. An argument exactly parallel to the one just given shows that a local hidden variable theory cannot reproduce the predictions of quantum mechanics concerning combinations of such measurements.

For the moment, the points I want to emphasize about this argument are just those I stressed with respect to Bell's Theorem. First, the argument does not rule out hidden variable theories altogether, but only *local* hidden variable theories. Second, it only rules out local hidden variable theories to the extent that it rules out locality itself. In other words, its objection is not to the hidden variables as such, but to the locality.[18]

In chapter 9 we'll see how the GHZ argument, like Bell's Theorem, depends on the assumption that there is no advanced action. Given that we've seen that advanced action in the microworld is independently plausible on symmetry grounds, this assumption seems tendentious, to say the least! For the moment, however, I want to emphasize that it is not just supporters of the incompleteness interpretation of quantum mechanics who stand to benefit. Nonlocality is a hook on which both sides of the debate about completeness seem currently impaled.[19]

In sum, then, the main objections to the ignorance interpretation or hidden variable view of quantum mechanics are those provided by the no hidden variable theorems and Bell's Theorem.[20] Within the confines of the conventional debate—that is, in particular, with no consideration given to the possibility of advanced action—these arguments establish that a hidden variable theory would need to be both contextual and nonlocal, in order to reproduce the predictions of quantum mechanics. Most commentators seem to conclude that this is too high a price to pay. This seems a rather precipitous judgment, however, given that the main alternative view seems no more inviting.

WHAT IF THERE IS NO COLLAPSE?

Before we conclude this overview of the debate about the interpretation of quantum mechanics, there is one further major position which deserves to be mentioned. Recall that apart from nonlocality, the major difficulty facing complete description accounts of the state function is the measurement problem—the problem as to when the state function collapses. One tradition in quantum mechanics seeks to evade this problem by denying that the state function does collapse. Roughly speaking, the suggestion is that the state function appears to collapse only because we view reality from a less than all-encompassing perspective. In the case of Schrödinger's Cat, for example, reality does indeed involve both a live cat and a dead cat, even after we open the box. It also involves a version of ourselves who sees a dead cat and a version who sees a live cat. We are not aware of both versions, however, for we look at things from the perspective of just one of them. From our perspective, then, the cat looks definitely either alive or dead, and its state function appears to have collapsed. From the Archimedean stand-point, however, reality contains both components. The state function always evolves in accordance with Schrödinger's Equation.

I shall call this the *no collapse* interpretation.[21] Its best-known version is the so-called *many worlds* interpretation, which is often represented as the view that reality "splits" in measurement interactions, into component branches reflecting the different possible outcomes. The noncollapsed state function is then supposed to describe the complete collection of branches. However, it is far from clear that the notion of splitting is essential or helpful to the formulation of the view. At any rate, a less contentious approach seems to be to stress the perspectival character of our ordinary viewpoint. A better term is the *relative state* view, which does stress the perspectival nature of the account.[22]

The no collapse view is especially popular among physicists seeking to apply quantum mechanics to cosmology. This preference turns on an apparent difficulty which confronts the orthodox complete description view in this case, a difficulty which is really a corollary of the measurement problem. If the state function of a system only collapses when the system in question interacts with a measuring device, then it would seem that the state function of the universe as whole could never collapse, for the simple reason that by definition, the universe as whole never interacts with anything external to it. Accordingly, quantum cosmologists have become enthusiastic supporters of the no collapse view.

Philosophers have been rather less enthusiastic, and there are a number of astute critical studies of the no collapse proposal in the philosophical literature on quantum mechanics.[23] This critical project is complicated by the fact that the view is ill-defined in a number of crucial respects (among them the issue as to what, if anything, actually "splits"), and I shall not attempt to cover the same ground as these studies here. I want to mention just two apparent difficulties for the view. One of these is well recognized by the view's proponents. It is the issue as to why the many-branched reality it envisages should appear classical—in other words, as to why we should not be aware of any of the vast plurality of alternatives the view requires to be equally real. I shall come back to this point, but I want first to draw attention to a problem which seems not to be recognized at all by proponents of the no collapse view, and to be underrated even by its philosophical opponents. It concerns the interpretation of the notion of probability in a model of this kind.[24]

What can it mean to say that a given outcome has a certain probability, if all possible outcomes are guaranteed to occur? The usual approach is to try to characterize quantum probabilities in terms of proportions of the total number of branches. To say that Schrödinger's Cat has a 50 percent chance of survival, for example, is to say that it survives in 50 percent of the large number of equally real branches or histories which arise when the experiment is performed. Where the number of branches is infinite, the notion of proportion will need to be replaced by some suitably generalized measure.[25] Either way, however, the suggestion runs foul of an old point in the theory of probability. The classical objection to the attempt to analyze objective probabilities in terms of proportions of alternatives is that we need to assume that the alternatives are *equiprobable*, which makes the intended analysis of probability circular.

Consider a coin tossed repeatedly, for example. If we represent the possible outcomes by forks, then each toss produces two branches, and repeated tossing contributes to an ever wider tree of possible sequences. In the limit as the number of tosses goes to infinity, almost all branches are such that the proportion of heads tends to 50 percent. But it is a mistake to take this limiting frequency to represent the probability that the coin will land heads when tossed. After all, the tree takes exactly the same form if we grow it from a biased coin, in which case the limit frequencies don't tend to the real probabilities. The limiting frequencies only give the right answer if we assume that at each branching, all possibilities are equally probable. Since probabilities of mutually exclusive possibilities are additive, this amounts to assuming exactly what we wanted to derive, I think.[26] The frequencies displayed in the tree only correspond to the objective probabilities provided by quantum mechanics if we assume that branching is governed by these probabilities. So the frequencies cannot provide an acceptable *analysis* of the notion of probability in this context. If we were concerned about what probability might *mean* in a world in which all possible outcomes are actualized, we won't find an answer by appealing to frequencies.

Partly in virtue of this classical objection to the view that probability is to be understood in terms of ratios of frequencies of outcomes, contemporary philosophers who favor objectivist accounts of probability will typically analyze it, at least in part, in terms of its connections with *rational decision*. Probability is seen as an objective measure of rational degrees of confidence, or something similar.[27] No collapse models seem to have some unusual difficulties at this point, however. The principle that all possible outcomes are actually realized has profound consequences for our ordinary views about rational choice. A graphic way to make this point is in terms of what Dieter Zeh has called quantum Russian roulette. Suppose that someone offers us a large amount of money to be Wigner's Friend—that is, to take the place of Schrödinger's Cat in a version of the experiment once suggested by the physicist Eugene Wigner. The experiment is to happen while we are asleep, so that we won't notice anything at all in the worlds or branches in which we "lose"—that is, in which we die. In worlds in which we live, however, we shall wake up very rich. According to no collapse views we are guaranteed to wake up in some branch, no matter how small the probability of the quantum safety switch which prevents our death. So from the subjective point of view it will be exactly as if the benefactor simply gave us the money, with no strings attached. All our future selves will be winners.

By the same reasoning it would seem to be rational for the entire human race to gamble its existence for any benefit whatsoever. For example, if we found a process which produced a useful energy output with odds of, say, one in a billion, and destroyed the universe in all other branches, it would still be perfectly rational to set up a power station to exploit it!

So it is very unclear what probability might mean for a no collapse view. However, one solution might be to say that objectively speaking, quantum mechanics does not describe probabilities at all, but simply gives *proportions* of outcomes. The question as to why these proportions of outcomes give us rational decision probabilities could then be left for later, and it would be possible to argue that sometimes they don't. All the same, it does seem that for better or worse these consequences expose the no collapse view to an unusual form of empirical verification. For suppose we play Russian roulette, quantum style, and find ourselves surviving long after the half-life predicted by orthodox views. This would be very good evidence that the no collapse view was correct. If we wanted to share our evidence with skeptical colleagues, we would need to ensure that they too participated in the game, of course—otherwise they would be left saying "I told you so" to our corpse in most of their surviving branches. (Alternatively, we might persuade each of our colleagues to participate in his or her own private game, and point out that the view predicts that each will find that he or she does very much better than average—indeed, that he or she survives everybody else in the initial group!)

Before we begin to take these possibilities too seriously, however, it needs to be emphasized that the no collapse view is a sport of the complete description interpretation of quantum mechanics. It is a response *within the confines of that interpretation* to the apparent intractability of the quantum measurement problem. To the extent that there is a viable alternative to the complete description view, therefore, the no collapse proposal is a response to a nonexistent problem. I have argued that even within the confines of the conventional debate—even if we neglect advanced action, in other words—the case against the hidden variable approach is greatly overrated. So funding for quantum suicide missions would be a little premature!

MANY MINDS?

Supporters of the no collapse approach sometimes claim that it avoids the problem of nonlocality in quantum mechanics. This claim rests on the thought that the nonlocal correlations revealed by Bell's argument are correlations between measurement outcomes. The correlations arise when measurements are made on each of the two parts of a suitable system. Each measurement individually produces one of a range of possible outcomes, and the problem on nonlocality stems from the fact that these individual outcomes turn out to be correlated in a puzzling way.

How does the no collapse view avoid the problem? Simply by denying that the individual outcomes ever actually happen, in the right sort of way. From the external standpoint, on this view, the measurements concerned simply don't have definite outcomes. They seem to have outcomes from the internal standpoint, of course, but there are as many internal standpoints as there are possible measurement results, and none of them is privileged— none provides the "true" result of measurement.

Thus the Bell correlations do not exist objectively, on this view. They only exist from the standpoint of a particular observer, who has gathered data from the remote measuring devices. Once gathered in this way, the elements of the correlation are no longer nonlocal: they exist side by side in the record book of a single observer, in a particular branch of the great tree of worlds. The suggestion is that the mistake of taking these correlations to be problematic arises from the mistake of thinking that the data records a single objectively existing reality, which predates the formation of the record in a particular branch.

A particularly striking form of this kind of defense of locality rests on the idea that the individual branches are entirely mental in character, and have no independent physical status. This is one feature of a so-called many minds interpretation of quantum theory proposed by the philosophers David Albert and Barry Loewer. Albert describes the resulting response to Bell's Theorem like this:

> What Bell proved is that there can't be any local way of accounting for the observed correlations between the outcomes of measurements like that; but of course (and this is the crux of the whole business) the idea that there ever *are* matters of fact about the "outcomes" of a pair of measurements like that is just what *this* sort of picture *denies!*[28]

As Albert and Loewer recognize, this is a very expensive solution to the problem of nonlocality. It requires not only the independent existence of mind and matter, and thus a radically dualist picture of reality, but also, as Albert explains, "that every sentient physical system there is is associated not with a single mind but rather with a *continuous infinity* of minds."[29] Albert also points out that it has disturbing epistemological implications: our present experience is not "incompatible with the hypothesis that the quantum state of the universe is (for now and for all time) [the] *vacuum* state,"[30] for example.

And for all this, what does the many minds suggestion actually buy? It relegates nonlocality to the world of mental appearances, but the fact remains that quantum mechanics is different from classical physics, on this view, in that classical physics does not give rise even to nonlocal *appearances*. This means that someone who sees the principle of no action at a distance as a prescription on what is physically acceptable *in the phenomena*—in what we *observe*—is not going to be satisfied with this solution. And why shouldn't we read the principle this way, if we are going to see things in these dualist terms? (After all, classical phenomenalists will read the principle this way as a matter of course.)

So the many minds solution to the problem of nonlocality turns on a reading of the locality constraint which the proposal itself gives us every reason to challenge. In a less dramatic way, the same seems to be true of the claims of other versions of the no collapse view to avoid nonlocality. All these claims turn on the idea that the Bell correlations are not fully objective—not part of an observer-independent physical world. But the fact remains that in classical physics we find no such correlations, even in the world of appearances. So it is something of a misrepresentation to say that these views make quantum mechanics local. The appearances remain nonlocal and non-classical, and the views concerned simply propose an explanation of these appearances, in terms of a theory which is not itself nonlocal. It would be better to say that they explain nonlocality than that they eliminate it.

With this made explicit, we can begin to compare this approach to others that might be on offer. I want to argue that advanced action provides the prospect of a much more economical and well-motivated explanation of the Bell correlations—indeed, an explanation which shows that they are not really nonlocal at all, in that they depend on purely *local* interactions between particles and measuring devices concerned. (They *seem* nonlocal only if we overlook the present relevance of future interactions.) In comparison to the no collapse views, the great advantage of this approach—apart from the fact that it is independently motivated by symmetry considerations—is that it doesn't require any radical separation between the real world and the world of appearances. It doesn't inter us in the shadow world of Plato's cave, forever remote from the vast reality of the never-collapsing state function.

Anyone who is attracted to the no collapse views by the promise of a solution to the problem of nonlocality would thus be well advised to be rather cautious. For one thing, the cure may be far worse than the disease. Is nonlocality really so implausible that the metaphysical apparatus of the many worlds view, let alone the many minds view, is preferable to a modest nonlocal hidden variable theory such as Bohm's? Bell himself thought not, and Einstein might well have agreed with him. More important, however, it may be that there is a far less drastic cure to hand. Advanced action offers to explain the Bell correlations, without genuine nonlocality, and without any of the metaphysical complexity of the many world view.

Before we turn to the advanced action approach to quantum mechanics, I want to mention a prominent aspect of contemporary no collapse theories which deserves to be considered in its own terms—in part, I think, because it seems to have an important role to play in quantum theory even if the no collapse view itself is rejected.

THE DECOHERENCE APPROACH

I noted above that one of the problems which is thought to confront the no collapse view is that of explaining why the macroscopic world normally seems classical, given that it actually comprises a vast number of distinct branches, reflecting the quantum mechanical apparatus of superposition, interference, and so on. This problem has been addressed by a proposal interestingly similar to an idea we encountered in chapter 2. We saw there that there is a tradition in statistical mechanics which seeks to justify Boltzmann's assumption of molecular chaos, and hence provide secure foundations for the *H*-theorem, in terms of the idea that real systems are never completely isolated from their environment. As Burbury put it, real systems seem always to be subject to outside influences "coming at haphazard," which (the argument runs) serve to defeat any correlations which might incline a system to become more ordered.

In the present context, however, the suggestion is that random external influences explain why we don't see quantum effects on a macroscopic scale, and perhaps more important still, why we see them rather selectively on the microscopic scale.[31] The basic idea may be appreciated by reflecting on the EPR argument. We noted that the strange entanglement between the two particles in an EPR experiment persists only so long as they remain unmeasured. Once we make a measurement on either particle the "coherence" is lost, and the particles behave as individual systems. The idea of the decoherence program is that in most cases this kind of effect takes place in a quite uncontrollable way, due to the influences of the external environment over which we have little effective control. The environment often acts as a measuring device, in effect. Uncontrollable interaction with the environment thus makes quantum systems behave like classical mixtures, and explains for example why we can't conduct a Bell experiment with nonmicroscopic components.

This proposal needs to be developed with some care, however. One source of confusion is that proponents of the decoherence approach sometimes talk as if decoherence can be regarded as an answer to the measurement problem within the confines of an orthodox collapse interpretation. Thus the physicist W. Zurek claims that "decoherence … can supply a definition of the branches in Everett's [no collapse] ... interpretation, but it can also delineate the border that is so central to Bohr's [collapse] point of view."[32] I think this is a mistake, however. There is a world of difference between saying "the environment explains why collapse happens where it does" and saying "the environment explains why collapse *seems* to happen, even though it doesn't *really* happen"—and the second thing is what the decoherence view should be saying. After all, the former suggestion does nothing to solve the problem of the universe as whole: if *its* state function is to collapse, there is no "environment" to do the job!

This is a graphic way of making the point that if quantum mechanics is to claim that Schrödinger's Equation applies—that the state function doesn't collapse—whenever physical systems are left to their own devices, it is difficult to see how the uncontrollable or complex nature of some real systems could make any difference. Again, it would be easy to construct what philosophers call a slippery slope argument—the kind of argument involved in our generalization of the Schrödinger's Cat experiment. In this case we need to start with a system which is agreed not to collapse, and to imagine adding complexity or environmental influence a particle at a time. It seems quite implausible that any one of these additions could suddenly break the wave function's back, and yet this is what the claim that the effects of the environment *actually* collapses the wave function seems to require.

Let us assume therefore that the decoherence view intends to say the second thing, viz., that the state function never collapses, but that environmental decoherence explains why it *seems* to collapse. This is an attractive idea, but one that seems to me to be undermotivated. What is decoherence actually buying, if not something explained in terms of *actual* (environmentally induced) collapse of the wave function? Zurek suggests that without decoherence our own brains might be in quantum superpositions, and that this would make us "conscious of quantum goings-on."[33] But who is to say what being in a quantum superposition would feel like? Not the supporters of decoherence, who think that it never happens!

Any version of the no collapse view is committed to the claim that in some sense reality contains equally real branches corresponding to "seeing the live cat" and "seeing the dead cat." Why isn't reality of this kind experienced as something ambiguous or fuzzy? Because we only experience one branch at a time—at least, any version of a no collapse view has to say something like this. But it is not clear what difference it makes to the plausibility of this claim whether the state function that describes the collection of branches has the form of a "genuine" pure state—a typical quantum superposition—rather than the kind of ersatz approximation to a classical mixture that decoherence is supposed to provide.

Note that it is not a real mixture, for the probabilities do not reflect uncertainty as to which of the possible alternatives is the actual one. All possibilities are equally real, on this view, and the effect of decoherence is just to reduce the interference terms characteristic of quantum pure states to very low levels. Elsewhere in the debate about quantum mechanics the difference between pure states and mixtures is important because pure states make things harder for ignorance interpretations, but this not an issue here. In this context we are taking for granted that all the branches are equally real. (Perhaps the claim is that if it weren't for decoherence we would experience the effects of the interference terms. Again, however, this seems to rely on hypotheses about the nature and physical basis of experience whose justification does not seem to have been provided.)

A related contribution claimed for decoherence is that it solves the so-called preferred basis problem. This problem arises because superpositions can be generally be expressed in many different ways. A given state function can be represented as a linear sum of many different sets of basic components. (Here's a close analogy. When we represent positions on a map by grid references, there is no geographical law that commits us to the usual north-south-east-west grid. We could choose our grid lines in any pair of perpendicular directions, and easily convert references based on one grid to references based on another. Quantum state functions can be rewritten in much the same way.) If we experienced components of raw superpositions, which of many possible components would we experience? The decoherence view argues that the effects of interaction with the environment privilege certain representations over others, and hence provide the required preferred basis.

This seems to be what Zurek means by saying the decoherence approach "supplies a definition of the branches in Everett's ... interpretation."[34] This problem seems a bit sharper than the previous one, but even here it is difficult to put one's finger on the precise issue. The problem tends to slide back into an issue about why we experience things in the way we do, and again I think that we haven't been told how things would seem if it weren't for decoherence. It is true that the issue takes a sharper form in some many world formulations of the no collapse views. If the world *really* splits, then we need a preferred basis to tell it where to split. But it is not clear that more plausible views closer to Everett's relative state view need anything like this. The view might be that reality can be carved in many ways—how you do it depends on "where you stand."

To sum up, my impression is that when properly located within the context of a no collapse view—that is, when detached from the claim that environmental effects *actually* collapse the wave function—decoherence is commonly supposed to do two closely related jobs. It is supposed to give superpositions a preferred basis, and in the process it is supposed to explain why reality seems classical. What I have trouble seeing is why someone who accepts the no collapse view, and therefore already accepts that experience gives us only a very narrow and perspectival view of reality, should regard these as serious problems.

However, I want to emphasize that these remarks concern the use that no collapse theorists claim to make of decoherence, not the decoherence program itself. Indeed, although most of the advocates of decoherence are no collapse theorists, this seems to be quite inessential. For according to one important way to understand of the decoherence program, it has nothing in particular to do with the no collapse approach. On this view, the program simply "starts with the formalism of quantum theory and studies the question of when classical equations of motion are good approximations to quantum theory, and in particular when (and to what extent) Bell-type effects can be neglected."[35] Issues of this kind arise on any interpretation of quantum mechanics. Someone who favors hidden variables will still need to explain why the peculiarly quantum statistics seem to apply in some cases but not others—why Bell correlations show up only under certain circumstances, for example—and an explanation in terms of practically uncontrollable coupling with the environment might very well work in this framework.[36]

The fact that these issues arise for any view of quantum theory helps to confirm the impression that the present association between the decoherence program and no collapse approaches is confused and poorly motivated. Whatever the distinctive problems of the no collapse view, they can't be addressed by a tool which is made for a different job. At present, then, the no collapse view seems to be gaining credence it hasn't earned from the promise of decoherence; while the decoherence program gains a reputation it doesn't deserve from its association with the excesses of the no collapse view. The association needs to be broken, if only to ensure that decoherence may take its place as an ingredient of other approaches to quantum mechanics. If advanced action is to explain the peculiarities of quantum statistics, for example, then we shall need an explanation as to why its effects do not show up more generally. The effects of environmental coupling might well be the key to this account.

SUMMARY: EINSTEIN'S LIVE ISSUE

In this chapter I set out to provide a very broad overview of the contemporary debate about the meaning of quantum mechanics, as it appears if the possibility of advanced action is (as usual) completely neglected. We have seen that there is a great division in the field, on an issue identified by Einstein early in the history of the theory. On the one side are those people who take quantum theory to provide a complete picture of a fuzzy reality—those who deny that there is any further aspect to physical reality, not characterized by the quantum state function. This was the position of Bohr and the Copenhagen school, and it quickly became the dominant view. On the other side are those who take the state function to provide an incomplete description, a kind of fuzzy picture of a sharp reality. This was the view of Schrödinger and Einstein himself, and of Bohm and Bell a generation later.

It is fair to say that the case in favor of each of these positions consists mainly of the arguments against the rival view. From a contemporary standpoint these arguments may be summarized as follows.

Against the complete description view:

• Schrödinger's Cat. Schrödinger's original argument was that the complete description view of the state function in microphysics would also commit us to real macroscopic superpositions—of a live and dead cat, for example—which seems absurd. However, the standard response to this objection is that the state function simply "collapses" before we get to this macroscopic stage.

• The measurement problem. By considering a range of variations of the Schrödinger's Cat experiment, in which the cat is replaced by a variety of different "measuring devices," we drew attention to the issue as to where the collapse takes place. This is a profound and unresolved difficulty for the complete description view.

• The EPR argument. We noted that the original EPR argument for the incompleteness of quantum mechanics is undermined by Bell's work, which shows that the EPR assumption of locality is untenable. However, even the framework of the original argument is sufficient to show that the complete description view is itself committed to nonlocality, in a way which appears particularly problematic in the light of special relativity. It requires a preferred frame in which collapse is instantaneous, for example. It is also an objectionable feature of the complete description view that it requires us to reject outright the possibility of an *explanation* of the EPR correlations.

• The problem of time asymmetry. As standardly represented, collapse is time-asymmetric—it depends on the past but not the future, in a way which does not seem explicable either as a harmless conventional asymmetry or as of the same origins as the thermodynamic asymmetry.

Against the incompleteness view (also known as the ignorance interpretation, or hidden variable view):

• The no hidden variable theorems. As exemplified by the shell game example, these seem to show that a hidden variable theorem could not possibly reproduce the predictions of quantum theory. However, they leave open the possibility of contextual hidden variable theories.

• Bell's Theorem and the GHZ argument. These show that even a contextual hidden variable theory would need to be nonlocal.

As things stand, then, both sides are troubled by nonlocality, and so it is rather puzzling that the hidden variable approach has not been more popular. By and large, commentators dissatisfied with the orthodoxy have preferred to move in the opposite direction. In order to avoid the measurement problem they embrace the no collapse view, rather than moving back to a more Einsteinian view of quantum theory. The no collapse view brings new problems, among them an issue about the meaning of probability in quantum mechanics.

In the end, however, the main objection to the no collapse view is that there are much more economical approaches to the problems it claims to address. I think that the popularity of this approach in contemporary physics can only be explained by the fact that the complete description interpretation had already become the orthodoxy. Having already turned their backs on Einstein's alternative, physicists who appreciated the untenability of the orthodox position saw no option but to press on into deeper water. The choice seems to me a very questionable one, even by the standards of the conventional debate. Contemporary physicists would do well to pay more attention to the cautious voice of the century's great revolutionary.

As I promised at the beginning of the chapter, then, this survey has turned out to favor Einstein's view of Einstein's issue. This recommendation is soon to be overshadowed by another, however. In the next chapter I want to argue that when the possibility of advanced action in quantum mechanics is given the attention it deserves, the case for the incompleteness view becomes even more appealing. In a very decisive way, advanced action seems to favor Einstein's view of the significance of quantum theory. Curiously, it also vindicates one of Bohr's main contentions—it entails that what we find in reality is in part a product of the fact that we have looked. However, it achieves this in such a subtle way that the two great antagonists of quantum theory end up arm-in-arm.

9

←——————→

The Case for Advanced Action

Some of the most profound puzzles of quantum mechanics are those that stem from the work of John Bell in the mid-1960s. As Bell's work became famous, he was often asked to survey the state of the subject, especially in light of his own contribution. He would typically conclude such a lecture by listing what he saw as possible responses to the difficulties, indicating in each case what he took to be the important physical or philosophical objections to the response concerned. His intuitions in this respect were unfashionably realist—like Einstein and Schrödinger, he disliked the common view that quantum mechanics requires us to abandon the idea of a world existing independently of our observations. He therefore appreciated the irony in the fact that from this realist standpoint, his own work seemed to indicate that there are objective nonlocal connections in the world, in violation of the spirit of Einstein's theory of special relativity. As Bell puts it in one such discussion,

> the cheapest resolution is something like going back to relativity as it was before Einstein, when people like Lorentz and Poincaré thought that there was an aether—a preferred frame of reference—but that our measuring instruments were distorted by motion in such a way that we could not detect motion through the aether. Now, in that way you can imagine that there is a preferred frame of reference, and in this preferred frame of reference things do go faster than light.[1]

Bell reached this conclusion with considerable regret, of course, and would often note that there is one way to save locality, and avoid this conflict with special relativity. Bell's Theorem requires the assumption that the properties of a quantum system are independent of the nature of any measurements that might be made on that system in the future—"hidden variables are independent of later measurement settings," to put it in the jargon.

Bell saw that in principle quantum mechanics could be both realist, in Einstein's sense, and local and hence special relativity friendly, by giving up this *independence assumption*. But he found this solution even less attractive than that of challenging special relativity, for he took it to entail that there could be no free will. As he puts it, in the analysis leading to Bell's Theorem,

> it is assumed that free will is genuine, and as a result of that one finds that the intervention of the experimenter at one point has to have consequences at a remote point, in a way that influences restricted by the finite velocity of light would not permit. If the experimenter is not free to make this intervention, if that also is determined in advance, the difficulty disappears.[2]

Physicists and philosophers have paid very little attention to these remarks. Philosophers, in particular, might have been expected to show more interest. In effect, Bell is saying that nature has offered us a metaphysical choice of an almost Faustian character. We can enjoy the metaphysical good life in quantum mechanics, keeping realism, locality, and special relativity— but only so long as we are prepared to surrender our belief in free will! If Bell is right about this, the philosophical fascination of the case is hardly diminished by the fact that he himself chose to resist the temptation.

Indeed, the choice would be fascinating enough if we were merely spectators, an audience to Bell's Faust. As it is, of course, the same offer is extended to all of us, and many of us might feel that Bell was wrong to refuse. After all, many thinkers have long since concluded that there is no such thing as free will, and might thus take the view that nature is offering a very attractive free lunch. Others might feel that the history of philosophy has taught us how to juggle free will (or some acceptable substitute) and determinism, and hence that we might hope to take advantage of nature's offer at very little real cost. (With respect, after all, who is Bell to teach philosophy what is incompatible with free will?) Even if Bell is right, and it does eventually come down to a choice between a relativistically acceptable realism and free will, we might feel that Bell simply makes the wrong choice—what we should say, as honest empiricists, is simply that science has revealed that we have no free will.

At any rate, one of my aims in this chapter is simply to bring this fascinating issue to a wider audience. However, I also want to argue that the offer is a much better one than Bell himself believed. I want to show that we may help ourselves to the advantages of giving up Bell's independence assumption—locality and Einsteinian realism—but save free will. The secret lies in advanced action. I want to show that the apparent problem about free will stems from two sources. First, the few physicists who have thought about giving up the independence assumption have tended to think not of advanced action but of another possibility, which does raise legitimate concerns about free will. Second, once the advanced action framework is made explicit, the remaining concern about free will arises from the mistake of reading the connection between the prior state of the physical system and the experimenter's choice in the wrong direction. Instead of taking the former to constrain the latter, we may reasonably take things the other way around, so that the experimenter remains in control.

In one sense, then, the advanced action proposal exploits the very same mathematical loophole that Bell himself is referring to, in the kind of remarks quoted above. While the mathematics remains the same, however, advanced action gives the idea a very different metaphysical gloss, and thus avoids what Bell himself saw as the loophole's disadvantages. The popular perception is that Bell is the man who proved that Einstein was wrong about quantum mechanics. I want to show, on the contrary, how close he came to being the man who proved that Einstein was right. I think it is fair to say that Bell saw how Einstein could be right about quantum mechanics, but didn't understand what he saw.

OUTLINE OF THE CHAPTER

Some of the argument in this chapter is quite dense, and as I don't know how to make it any easier, I want to begin with a map of the territory we are going to cover. As I mentioned two paragraphs back, one confusing factor is that there are actually two quite different ways of relaxing Bell's independence assumption. One way is to take the required correlation between hidden variables and future measurement settings to be established by some factor—a "common cause"—in their common past. The other is to take it to obtain simply in virtue of the *future* interaction between the system and measuring device in question. These two approaches agree on the core mathematics involved—on the nature of the correlation between hidden states and measurement settings—but disagree about the explanation of this correlation.

I want to defend the latter strategy. I think that its advantages have been overlooked either because the two strategies have not been clearly distinguished, or because the former strategy has seemed more plausible, in not countenancing backward causation. Bell himself seems to have been aware of both versions, and to have regarded both as incompatible with free will, but it is doubtful whether he saw them as clearly distinct. In my view the former strategy is objectionable on grounds that have nothing to do with free will, namely that it calls for a vast and all-pervasive substructure in reality to provide the required common causes. The latter view in contrast is elegant and economical, as compatible as need be with free will, and—as we saw in chapter 7—appears to respect a temporal symmetry which other views ignore.

The first task of the chapter will thus be to distinguish these two strategies for saving locality in the face of the Bell's Theorem. Having set aside the common cause proposal, I'll then focus on the advanced action strategy. I'll use the Ypiarian parable from chapter 8 to explain how advanced action accounts for Bell's correlations, and how it avoids both the standard objections to backward causation and Bell's own concerns about free will. The upshot seems to be that we may avail ourselves of Bell's route to a local realism about quantum mechanics, provided that we are prepared to recognize that quantum mechanics reveals the presence of advanced action in the world. This move does not have the disadvantages that Bell thought it had, but does have the major advantage foreshadowed in chapters 5 and 7: it seems to be precisely what temporal symmetry should have led us to expect in microphysics.

Bell's Theorem is not the only problem for hidden variables in quantum mechanics. As I explained in chapter 8, the other major objections stem from the so-called no hidden variable theorems, and from the new GHZ argument (which lies somewhere between the no hidden variable theorems and Bell's Theorem). I want to show that advanced action also promises to meet these objections. Concerning the GHZ argument, we shall see that advanced action exploits much the same loophole as it does with respect to Bell's Theorem itself. Concerning the older no hidden variable theorems, I shall offer an informal argument that the puzzling consequences of superposition in quantum mechanics are just the kind of phenomena we should expect in a world with advanced action—a world with time-symmetric dependence at the micro level. And I'll show how the *asymmetric* dependence we normally take for granted comes to be reflected in the basic assumptions of the no hidden variable theorems (and, to be fair, of most existing hidden variable theories themselves).

The chapter concludes with a brief discussion of some more general issues. Advanced action requires that we take seriously the atemporal Archimedean perspective in physics. But the ordinary temporal perspective is so familiar, and so deeply embedded, that we need to be suspicious of many of the concepts used in contemporary physics. We can't simply assume that familiar concepts will carry over smoothly to an Archimedean physics. The case of causation provides an object lesson: we learned in chapters 6 and 7 that it is no easy matter to separate the objective aspects of causality from those more subjective aspects which reflect our own asymmetric perspective. In the final section of this chapter I point out that other familiar notions in physics may also need careful consideration of this kind, if the new physics is not to fall for old mistakes.

LOCALITY, INDEPENDENCE, AND THE PRO-LIBERTY BELL

As we have seen, Bell's Theorem is standardly taken to show that quantum mechanics implies nonlocality. However, the theorem requires the assumption that the values of hidden variables are statistically independent of future measurement settings. While this independence assumption is generally taken for granted in the literature, Bell himself sometimes considered relaxing it, in order to defend locality. On balance he regarded such a move as even less attractive than nonlocality, however—this despite the fact that he himself believed that nonlocality conflicts with the orthodox interpretation of special relativity. Few other commentators pay any explicit attention to the possibility of giving up the independence assumption. I am not sure whether this is because Bell's concerns about freedom are widely shared, or simply because the independence assumption seems too obvious to challenge. At any rate, it is this unconventional path that we are now interested in exploring.

I think that one of the reasons why this path has remained unexplored is that there are two very different ways of relaxing the independence assumption. One of these is initially more obvious and more appealing than the other, but soon runs into seemingly insuperable obstacles. It is easy to mistake these obstacles for objections to the general proposal to drop the independence assumption, rather than simply to one way of implementing that proposal. It seems that people who have considered giving up the independence assumption have failed to notice that there is a less obvious but more attractive way of doing so. (As I'll explain, Bell himself seems to be one person who made this mistake.) We need to distinguish the two options, and set aside the more popular one, so that we can turn our attention to the less obvious, the more surprising, and yet (I want to argue) very much the more attractive of the two.

Here's the basic set-up. A hidden variable theory wants to say that any quantum system S has some underlying or hidden state in addition to its ordinary quantum state. (Call this hidden state λ.) Suppose that such a system S is approaching a measurement device M, which may be set in various ways. (If S is a photon, for example, and M a polarizing lens, then M may be rotated to any position in a 180° range, perpendicular to the line of flight of the photon.) The independence assumption says that the value of λ does not depend on the setting of M.

Let's think about what it would mean if the independence assumption failed. Suppose for simplicity that it fails in an extreme way, so that we have a perfect correlation between certain possible values of λ and certain measurement settings. In effect, then, we thus have a correlation between two spatially distant events or states of affairs: the current state of the incoming quantum system, on the one hand, and the current state of the measuring device, on the other (or perhaps the current precursors of that state, if the setting itself has not yet been made).

How could such a correlation arise? If we are contemplating this possibility as a means of avoiding nonlocality, then we are not going to be interested in the possibility that it might arise directly, as it were, via some kind of action at a distance. In other words, we'll make it a matter of stipulation that neither event acts on the other directly. We thus have a correlation between two events, but not a correlation which stems from the fact that one event is a cause of the other.

Reichenbach's principle of the common cause, which we encountered in chapters 5 and 6, tells us that in practice, correlations of this kind turn out to be explained by the fact that both events are correlated with an *earlier* common cause. Indeed, as we saw, Reichenbach's principle simply formalizes an idea that we take for granted in ordinary life. By these ordinary standards, then, it is natural to assume that the independence assumption could only fail if there were some factor in the common *past* of the quantum object and the measurement device, responsible for their correlation.

Let's call this the *common past hypothesis*. It is the first of what will turn out to be two quite distinct ways of relaxing the independence assumption. I'll come to the second way in a moment, but first let's think about the common past hypothesis, and whether it is a plausible way to try to save locality in quantum mechanics, in the face of Bell's Theorem.

LOCALITY SAVED IN THE PAST

The common past hypothesis seems to be what Bell himself usually had in mind, when he considered relaxing the independence assumption. The following passage is the most explicit comment on the matter I know of in his published work.

> It may be that it is not permissible to regard the experimental settings *a* and *b* in the analyzers as independent variables, as we did. We supposed them in particular to be independent of the supplementary variables λ, in that *a* and

b could be changed without changing the probability distribution $\rho(\lambda)$. Now even if we have arranged that *a* and *b* are generated by apparently random radioactive devices, or by the Swiss national lottery machines, or by elaborate computer programmes, or by apparently free willed physicists, or by some combination of all of these, we cannot be *sure* that *a* and *b* are not significantly influenced by the same factors λ that influence [the measurement results] *A* and *B*. But this way of arranging quantum mechanical correlations would be even more mind boggling than one in which causal chains go faster than light. Apparently separate parts of the world would be conspiratorially entangled, and our apparent free will would be entangled with them.[3]

As this passage suggests, Bell's main objection to the common past hypothesis seems to have been that the degree of determinism required seems incompatible with free will. He reinforces this impression in the published interview from which I quoted earlier in the chapter.

The free will objection is a little puzzling, however. After all, a long philosophical tradition, called "compatibilism," maintains that free will is compatible with classical Laplacian determinism. Surely the determinism involved here would be no more severe than that? (How could it be?) The compatibilists might be wrong, of course, but why prejudge the issue? More important, why take the issue to be relevant *here,* and not, say, as an objection to classical mechanics?

I suspect that Bell was influenced by two considerations. The first is a fatalist argument quite distinct from causal determinism: roughly, it is the thought that if the state of the incoming quantum object "already" reflects the measurement setting, then we are not free to choose that setting. The second factor is the thought that the required common cause would have to be something entirely new, something of a kind not presently envisaged in our view of the physical world. The conclusion that our actions are influenced by physical states of affairs of a previously unimagined kind could easily lead one to fatalism!

I'll return to the first of these factors later on. As for the second, it suggests an argument against the common past hypothesis which has little or nothing to do with fatalism. This hypothesis requires a universal mechanism of quite extraordinary scope and discrimination, in order to maintain the required correlations. Think of all the different ways in which measurement settings might be chosen. The mechanism would need to steer all these different physical pathways toward the same endpoint.

This objection could be made more precise, of course, but the basic point is clear enough. The common past hypothesis needs to postulate a vast hidden substructure underlying what we presently think of as physical reality, in order to supply the required common causes. If this is what we need to save locality, then the cure seems worse than the disease.

LOCALITY SAVED IN THE FUTURE

Suppose then that we reject the common past hypothesis. Is there any other way in which the independence assumption might fail? I think there is. In terms of my idealized example, there might simply be a brute correlation between the present values of λ (the hidden state of the quantum object) and the future values of the measurement setting—a correlation explicable not in terms of some extra common cause in the *past*, but simply in terms of the existing interaction in the *future*.

Let's call this the *common future hypothesis*. It suggests that the pattern of correlations has the structure of a ∧-shaped fork. The vertex of the fork is the point at which the object and the measuring device interact in the future, and the two prongs represent the present values of λ and the measurement setting.

It may be helpful to illustrate the difference between this and the common past hypothesis in terms of the Ypiarian parable from chapter 8. Recall the salient points: Ypiarian twins were interrogated separately, and each asked one of three questions. When they were asked the same question, they always gave the same answer. When they were asked different questions, they gave the same answer 25 percent of the time. We saw how the first of these results seemed to imply that there was some common factor, some characteristic that pairs of twins always shared, responsible for their answers; whereas the second result seemed to conflict with this conclusion, in the sense that the assumption of underlying factors of this kind enabled us to predict that when different questions were asked, we should have expected the same answer at least 33 percent of the time. This conflict between the two kinds of result was what we called Doppelgänger's Twin Paradox, for Doppelgänger's great contribution was to uncover the puzzle that lay hidden in these statistics. Let us take up the story.

Ypiaria 3: the interrogators strike back

Very little is known, of course, about the factors which must have governed an Ypiarian's answers to the three questions permitted under Ypiarian law. The surviving records tell us what was said but not in general why it was said. The psychological variables are hidden from us, and must be inferred, if at all, from the behavioral data to which we have access. However, the puzzling character of the data is easily explained if we allow that the relevant variables are not independent of the "measurement settings"—that is, of the choices the interrogators make about which of the three available questions they will ask on a given occasion.

As in chapter 8, let us denote the underlying psychological state of the twins by expressions such as NYY, YNY, and so on, indicating what answer they would give to each of the three possible questions. We saw that the results of cases in which both twins are asked the same question suggest that Ypiarian twins are always identical to each other with respect to these underlying states. This inference itself depends on the independence assumption, for on any given occasion on which the same question is asked of both twins, we only discover the twins' dispositions with respect to one of the three possible questions. In order to infer that the twins concerned have the same disposition with respect to any of the three questions, we need to assume that the disposition actually revealed *would have been present* even if a different question had been asked—in other words, that it doesn't depend on the choice of measurement. For the moment, however, let us continue to assume that Ypiarian twins do share dispositions of this kind. I want to show that even granting the conclusion drawn from this use of the independence assumption, Doppelgänger's results are easily explained if we are prepared to abandon the independence assumption itself at a later stage. At least for the sake of illustration, in other words, we can put all the weight on the explanation of the second component of the Twin Paradox, the case in which the twins are asked different questions (see 8·2).

Consider for example a pair of twins who are going to be asked questions 2 and 3, respectively. If the twins' underlying state is not independent of this choice of questions, we may postulate that the fact that the questions are different makes it less likely that the twins will be in the states YYY, YNN, NYY, and NNN that yield the same answer to questions 2 and 3; and correspondingly more likely that they will be in one of the states YYN, YNY, NYN, and NNY that yield different answers to these questions. At least in a purely mathematical sense, it thus becomes very easy indeed to explain why we find 25 percent rather than 33 percent in 8·2.

This loophole requires that there be a correlation between the underlying state of the each twin, and the external state of the world which results in that twin being asked one question rather than another. Why do we ordinarily find this such an implausible possibility? Simply because we take it for granted that events in the world are independent, unless they share some common causal history. In effect, this is our old friend, the principle of μInnocence. With this assumption safely in place, the only way in which the state of each twin could be correlated with that of the interrogators would be by virtue of some connection in their common past—some common cause, responsible for establishing the required correlation.

Thus with μInnocence in place, we are led to the common past hypothesis as the only way in which the independence assumption might fail. As Bell himself emphasizes in the passage quoted above, however, there is much we can do to reduce the plausibility of this hypothesis. In particular we could ensure that the choice of measurement settings was made by some elaborate randomizing device, so as apparently to obliterate all traces of influence from anything that might lie in the causal history of the underlying states we wish to measure. (Bell suggests that champion of chance, the Swiss national lottery machine.) Another possible strategy would be to allow the measurement settings to be determined by influences coming from regions of the universe too distant to have exerted any previous causal influence. In order to evade countermeasures of these kinds, a common cause would need to be something quite new: something so pervasive as to reduce the apparent causal order of the world to a kind of epiphenomenal shadow play (and very likely something which would itself be in conflict with the conceptual framework of special relativity). As I noted earlier, the threat to free will is far from the most serious objection to this way of attempting to motivate the proposal that we abandon the independence assumption.

But what of μInnocence itself, the unchallenged linchpin of this line of argument? In earlier chapters we explored the connection between μInnocence and the principle PI^3. I argued that to the extent that PI^3 holds—to the extent to which the world is temporally asymmetric in this way—this is a macroscopic feature of reality, a manifestation of the fact that we find statistically unlikely coherent organization of microscopic events toward our past but not apparently toward our future. It gives us no reason to accept the *microscopic* asymmetry embodied in μInnocence.

What does abandoning μInnocence do for us? In the Ypiarian case, it allows us to suggest that the required correlation between the underlying psychological states of each twin and the external "measurement settings"—the choice of questions—does not depend on anything in the common past of these different states of affairs. Rather, it depends on their common future, on the fact that the particular question comprises a part of the "future history," or *fate*, of the twin concerned.[4]

This, then, is the common future hypothesis. It holds that hidden variables are dependent on (correlated with) the fate of the twins concerned, as well on their history. It allows that at the time when the two twins separate, they already know, in effect, what questions they are going to be asked, and are therefore able to arrange their later answers in light of this information. Thus it becomes a very easy matter to ensure that on 25 percent of the occasions on which they going to be asked different questions, the twins adopt states which yield different answers to those questions.

In the case of quantum mechanics, one of the puzzles of this debate is that the common future hypothesis does not seem to have been considered explicitly by Bell himself. Bell devotes much more thought than most of his contemporaries to the possibility of abandoning the independence assumption, but then seems to think of this possibility in terms of the common past hypothesis. Why should this have been so? The next section offers some brief speculation on this historical point.

WAS BELL TOLD?

In one sense, Bell himself can hardly have been unaware of the common future hypothesis, as a way of giving up the independence assumption. The idea that Bell's work needs to be explained in terms of advanced action, or backward causation, is an old idea on the fringes of quantum mechanics. (Bell certainly knew of the early work of O. Costa de Beauregard, for example, in which Costa de Beauregard argues that the Bell correlations require what he calls zig-zag causality.)[5] However, I suspect that Bell did not appreciate that what this idea involves—the common future hypothesis—is very different from the common past hypothesis. My suspicion is based in part on his response when I wrote to him in 1988, sending two of my own early papers on advanced action in quantum mechanics. In his reply, he made it clear that he had already thought about such things: "I have not myself been able to make any sense of the notion of backward causation. When I try to think of it I lapse quickly into fatalism." But for his own thoughts on the subject, he then went on to refer me to a published discussion with Clauser, Horne, and Shimony, in which what is considered turns out to be the common past hypothesis rather than the common future hypothesis.[6] (There is no mention of backward causation in this discussion, for example.) At the time I thought of this as simply a slip on Bell's part, but it now seems to me to reflect the fact that he never properly distinguished these two alternatives.

I suspect that Bell's failure to draw this distinction can be traced in part to his concerns about fatalism. In effect, Bell thought that both approaches conflict with our intuitive assumption that experimenters are free to choose measurement settings, and that these settings are free variables. At one point in the discussion just mentioned, he characterises this assumption as the idea that "the values of such variables have implications only in their future light cones. They are in no sense a record of, and do not give information about, what has gone before." It is true that both hypotheses deny this: both take the setting of the measurement device to be correlated with the earlier state of the object system. But this is where the similarity ends. Because the two views tell very different stories about what sustains this correlation, they may have very different implications concerning human freedom. In the case of the common future hypothesis, as we shall see, it is plausible to argue that the relevant earlier states are under the control of the experimenter who chooses the measurement setting—and hence that the violation of Bell's condition (as quoted above) does not imply that the measurement settings are not free variables, in the most useful sense of that term.

In fact, as I noted earlier, there seem to be two quite separate arguments for fatalism in this debate: one turns on the thought that if the photon's state is "already" correlated with the future setting of the polarizer, then we can't really have any choice in the matter, the other on the idea that hidden common causes would be incompatible with free will. The first argument is a bad one, for reasons I'll explain in a moment. The second argument has more to be said for it, as we have already noted, but applies only to the common past hypothesis. It may be that Bell didn't appreciate that these are quite different arguments, and confused the two hypotheses as a result.

THE BENEFITS OF BACKWARD FORKS

Let me summarize the attractions of the common future hypothesis:

- Compared to the common past hypothesis, its major advantage is that it doesn't need to postulate a vast and universal mechanism to provide the required common causes.

- Compared to all other major approaches, its advantage seems to be that it does not conflict with special relativity. In the Ypiarian model, this is because the point at which twins become coupled—whether their conception, their birth, or some later meeting—lies well within the light cones of both their later interrogations. The effect is not instantaneous, not at a spacelike distance, and needs no mysterious carrier. It has the twins themselves, who bear the marks of their future as they bear the marks of their past.

- Compared to everything else on offer, in effect, it also has the advantage that it explicitly abandons a temporally asymmetric assumption—μInnocence— for which there seems to be no satisfactory justification. In other words, it respects an important symmetry which other approaches ignore.

With advantages like these, why has this approach not attracted more attention? A major factor, I think, is that the status of the common cause principle has not been well understood. This principle is very reliable elsewhere, and it is natural, other things being equal, to take it for granted in quantum mechanics. So physicists and philosophers who have contemplated abandoning Bell's independence assumption have tended to think in terms of the common past hypothesis. Only when we investigate the background to the common cause principle do we find that other things are not equal. On the contrary, it turns out we have good reason to expect the principle to fail in microphysics. But this point has not been noticed, by and large, and so the common future hypothesis has not been explored.

I don't think this is the only explanation for the neglect of the common future hypothesis, however. Resting as it does on a rejection of μInnocence, the hypothesis involves advanced action, or backward causation, in the sense we explored in chapter 7. And as I noted, the possibility of advanced action has sometimes been raised directly in quantum mechanics, without the confusing issue of the independence assumption. In particular, advanced action has sometimes been advocated on symmetry grounds—the guiding thought being that there is no reason, a priori, to expect causation to be temporally asymmetric. We have seen that this is an argument which needs to be handled with a great deal of care, however. The *ordinary* asymmetry of dependence turns out to be anthropocentric in origin, and therefore not to embody any violation of symmetry principles in physics. It is only when we inquire deep into the presuppositions of this ordinary practice that we find that it does embody an objective asymmetry if taken to apply without limit in microphysics. Without this underpinning, previous appeals to symmetry in support of advanced action in quantum mechanics have been misdirected, even if well intentioned.

This problem doesn't explain why the advanced action suggestions have been treated so harshly, however. Rather, most commentators think that there is something absurd or contradictory in the idea that the past might be correlated with the future in this way. This objection resolves into two main strands. The first is the suspicion that advanced action leads to fatalism— that it is incompatible with the ordinary supposition that the future events concerned are ones with respect to which we have independent present control and freedom of choice. The second is the thought that the correlations required for advanced action could be exploited to yield "causal paradoxes" of one kind or another. This claim depends on the venerable line of reasoning called the bilking argument, which we encountered in earlier chapters.

My view is that both these objections are vastly overrated. In order to show how the common future hypothesis escapes them, let's return to Ypiaria.

Ypiaria 4: causal paradox?
In chapter 7 we saw that the essence of the bilking argument is the thought that if there were backward causation (or time travel), it could be exploited to allow retroactive suicide and other less dramatic but equally absurd physical results. But is this true of the kind of backward influence postulated by the common future hypothesis in the Ypiarian case? Could Y.E.S., the

Ypiarian Euthanasia Society, hope to exploit these correlations in the interests of painless deaths for their aging members?

Regrettably not, I think. For what is the earlier effect—the claimed result of the later fact that a pair of twins T_{DUM} and T_{DEE} are asked questions 2 and 3, say? It is that the state of the twins when they last met is less likely to be one of those (YYY, YNN, NYY or NNN) in which questions 2 and 3 give the same response. For simplicity let's ignore the probability, and make the effect stronger: let's assume that being asked these questions *guarantees* that T_{DUM} and T_{DEE} will not be in one of these states at the earlier time. Retroactive suicide requires that this effect be wired to produce the desired result—that a device be constructed which kills the intended victim if the twins do not have one of the excluded states at the earlier time in question. (This machine might be too generous, if other future possibilities do not prevent the past state of affairs which triggers it; but this won't worry the members of Y.E.S., who won't be concerned that they might *already* have killed themselves—accidentally, as it were.) So the trigger needs to detect the relevant states of the twins, *before the occurrence of the claimed future influence on these states*—before T_{DUM} and T_{DEE} are next interrogated by the Ypiarian police. But why suppose it is physically possible to construct such a detector? We know these states can be detected by the process of interrogation—in effect they are dispositions to respond to such interrogation in a certain way—but this is not to say that they are ever revealed in any other way.

We thus have the prospect of an answer to the bilking objection. It exploits the loophole we discussed in chapter 7 (the one first identified by Michael Dummett). The answer points out that the thought experiment on which the bilking argument relies rests on an assumption that might simply be false: the experiment might simply be physically impossible. (This is a familiar response to arguments of this kind in science. For example, it was sometimes argued that space must be infinite, since if it were finite one could journey to the edge and extend one's arm. One response to this argument is to point out that even if space were finite the required journey might be physically impossible, because a finite space need have no edges.)

It is impossible at this distance to adjudicate on the Ypiarian case. Doppelgänger's work notwithstanding, we know too little of Ypiarian psychology to say whether the relevant states would have been detectable before interrogation. It is enough that because the bilking argument depends on this assumption, the advanced action proposal remains a live option.[7]

Ypiaria 5: the fatalist objection
Now to Doppelgänger's own main concern about the advanced action interpretation, namely that it seems to deny free will. If the Ypiarian interrogators'

choices of questions are correlated with the earlier psychological states of the twins concerned, then surely their apparent freedom of choice is illusory. For when they come to "decide" what questions to ask their "choice" is already fixed—already "written" in the mental states of their interviewees. This is the *fatalist* objection to advanced action.

The first thing to note about the fatalist objection is that it tends to slide back into a version of the bilking argument. If we think of "already determined" as implying "accessible," then we seem to have the basis of a paradox-generating thought experiment. Since we have already discussed the bilking argument, let us assume that "already determined" does not imply "accessible"—in other words, that something can be "already true," without necessarily being the kind of thing that we could find out about. What then remains of the fatalist point?

The strategic difficulty now is to set aside this objection to advanced action without a philosophical dissertation on the subject of free will. Two points seem to be in order. First, even if the argument were sound, it would not show that advanced action is physically impossible. Rather it would show that advanced action is physically incompatible with the existence of free will. In the interests of theoretical simplicity the appropriate conclusion might then be that there is no free will. Free will might then seem another piece of conceptual anthropocentrism, incompatible (as it turns out) with our best theories of the physical world. So the fatalist objection is strictly weaker than the causal paradox argument. If successful the latter shows that advanced action is impossible; the former, at best, merely that it is incompatible with free will.

The second thing to be said about the fatalist objection is that it has a much more familiar twin. This is the argument that if statements about our future actions are "already" either true or false (even if we can't know which), then we are not free to decide one way or the other. There are differing views of this argument among philosophers. The majority opinion seems to be that it is fallacious, but some think that the argument does rule out free will, and others that free will is only saved by denying truth values to statements about the future. On the last point, physicists perhaps have more reason than most to grant determinate truth values to statements about the future. Contemporary physics is usually held to favor the four-dimensional or block universe view of temporal metaphysics, whereas the thesis that future truth values are indeterminate is typically thought to require the rival tensed view of temporal reality.

If we accept that statements about the future have truth values, then there are two possibilities. If the ordinary fatalist argument is sound, then there is no free will. But in this case the fatalist argument against advanced action is beside the point. If there is no free will anyway, then it is not an objection to advanced action if it turns out to imply that there is none. If the ordinary argument is unsound, on the other hand, then so surely is the backward version. For the arguments are formally parallel. If the refutation of the forward version does not depend on what makes it the forward version—in effect, on the special status of future tensed statements—then it depends on what the two versions have in common. If one argument fails, then so does the other.

Indeed, what it seems plausible to say in the future case is something like this: statements about the future do have truth values, and some of these statements concern events or states of affairs which do stand in a relation of constraint, or dependence, with respect to certain of our present actions. However, what gives direction to the relation—what makes it appropriate to say that it is our actions that "fix" the remote events, rather than vice versa—is that the actions concerned *are* our actions, or products of our free choices. The fatalist's basic mistake is to fail to notice the degree to which our talk of (directed) dependence rides on the back of our conception of ourselves as free agents. Once noted, however, the point applies as much in reverse, in the special circumstances in which the bilking argument is blocked.

At any rate, the more basic point is that the fatalist argument usefully distinguishes the backward case only if two propositions hold: (1) the classical (future-directed) fatalist argument is valid, and thus sound if future tensed statements do have truth values; and (2) there is a significant difference between the past and the future, in that past tensed statements do have truth values, whereas future tensed statements do not. Proposition (1) is a rather unpopular position in a philosophical debate of great antiquity. Proposition (2) is perhaps a less uncommon position, but one that modern physicists seem to have more reason than most to reject. Taken together, as they need to be if the fatalist objection to the advanced action interpretation is not to collapse, these propositions thus form an unhealthy foundation for an objection to an otherwise promising scientific theory.

ADVANCED ACTION IN QUANTUM MECHANICS

As in the Ypiarian case, it is easy to explain Bell's results in quantum mechanics if we allow that particle states can be influenced by their future fate as well as their history. One way to do this would be to allow for a probabilistic "discoupling factor" which depended on the actual spin measurements to be performed on each particle and which influenced the underlying spin properties of the particles concerned. We might say for example that the production of such particle pairs is governed by the following constraint:

(9·1)   In those directions G and H (if any) in which the spins are going to be measured, the probability that the particles have opposite spins is $\cos^2(\alpha/2)$, where α is the angle between G and H.

Note that this condition refers to the *fate* of the particles concerned; it allows that their present properties are in part determined by the *future* conditions they are to encounter. Thus it explicitly violates Bell's independence assumption.[8]

How does 9·1 cope with the kind of objections to advanced action that we dealt with in the Ypiarian case? We saw that the causal paradox objection rested on the assumption that the claimed earlier effect could be detected in time to prevent the occurrence of its supposed later cause. What does this assumption amount to in the quantum mechanics case? Here the claimed earlier effect is the arrangement of spins in the directions G and H which are later to be measured. But what would it take to detect this arrangement in any particular case? It would take, clearly, a measurement of the spins of particles concerned in the directions G and H. However, such a measurement is precisely the kind of event which is being claimed to have this earlier effect. So there seems to be no way to set up the experiment whose contradictory results would constitute a causal paradox. By the time the earlier effect has been detected its later cause has already taken place. In effect, then, quantum mechanics thus builds in exactly what we need to exploit the loophole in the bilking argument.

The advanced action explanation of Bell's results thus lacks the major handicap with which it has usually been saddled. On the other side, it promises the advantages we noted earlier. For one thing, it does not seem to call for any new field or bearer of the influence that one measurement exerts on another. If we think of the fate of a particle as a property of that particle—a property which has a bearing on the results of its interaction with its twin—then the particles themselves "convey" the relevant influence to its common effect at the point at which they separate. More important, by thus confining the backward influence of future measurements to their past light cones, the advanced action account avoids action at a distance,[9] and hence the threat of conflict with Einstein's theory of special relativity.

The extent of the last advantage clearly depends on how much alternative explanations do conflict with special relativity. This has been a topic of considerable discussion in recent years.[10] It is widely agreed that the Bell correlations do not permit faster than light signaling, but the issue of causal influence is less straightforward. Whatever the nature of the influence, the concern seems in part to be that special relativity implies that any spacelike (i.e., faster than light) influence will be a backward influence from the point of view of some inertial frame of reference. Backward influence has seemed problematic on the grounds we have already examined: "It introduces great problems, paradoxes of causality and so on."[11] Now if this were the only problem with Bell's preferred nonlocal influences, it would weaken our case for preferring backward to spacelike influence. For it would mean that in rejecting the usual argument against backward causation we would also be removing the main obstacle to Bell's own faster than light interpretation of the phenomena.

There is another problem with Bell's view, however. A spacelike influence seems to distinguish one inertial frame of reference from all the rest. It picks out the unique frame according to which the influence concerned is instantaneous. This contradicts the accepted interpretation of special relativity, which is that all inertial frames are physically equivalent. It was this consequence that Bell was referring to in the passages I quoted at the beginning of the chapter. As he puts it elsewhere, the view commits us to saying that "there *are* influences going faster than light, even if we cannot control them for practical telegraphy. Einstein local causality fails, and we must live with this."[12] As I have emphasized, the advanced action idea requires no such reconstruction of special relativity. Unless there is more to be held against this approach than the empty charges of fatalism and causal paradox, then, it seems to offer a rather promising explanation of the Bell correlations!

EINSTEIN REISSUED?

Like Einstein, Bell had intuitions about quantum theory of a strongly realist character. He was attracted to the idea of an objective world, existing independently of human observers, and favored the view that quantum mechanics gives us only an incomplete description of this objective reality. (He liked David Bohm's hidden variable theory, for example.) As I have noted before, the extent to which Bell's Theorem conflicts with Einstein's view is commonly misrepresented. Even if we leave aside the loophole which turns on the independence assumption, Bell's result simply excludes *local* hidden variable theories—which, given that it appears to exclude almost all local interpretations, can hardly be counted a decisive objection to the hidden variable approach.

Without advanced action, then, Bell's Theorem is more or less neutral between Einstein's view and its more conventional rivals. With advanced action, however, the case seems to swing strongly in favor of the Einsteinian view. As it stands, quantum mechanics does not embody the backward influence which is required to avoid nonlocality. (On the contrary, as we have seen, the complete description interpretation of the standard theory involves the time-asymmetric pattern of dependency characteristic of μInnocence.) This means that *only* a model which adds something to the standard theory will be capable of representing advanced action. Only some version of the Einstein view seems able to save locality in this way.

Bell's Theorem aside, a revitalized Einsteinian view would have the attractions it always held for the interpretation of quantum mechanics. As we saw in the previous chapter, its great virtue is that because it denies that the collapse of the wave function corresponds to a real change in the physical system concerned, it does not encounter the measurement problem, which is essentially the problem of providing a principled answer to the question as to exactly when such changes take place. The fact that the measurement problem is an artifact of a particular way of interpreting quantum mechanics has tended to be forgotten as the views of Einstein's Copenhagen opponents have become the orthodoxy, but we saw that it was well appreciated in the early days of the theory. The point of Schrödinger's original use of his famous feline *gedankenexperiment* was to distinguish these two ways of looking at quantum theory—and to point out that unlike the view of the theory that he himself shared with Einstein, that of their Copenhagen opponents would be saddled with a major problem about the nature and timing of measurement.

Although it avoids the standard measurement problem, it might seem that the view I am suggesting—Einsteinian hidden variables with advanced action—will face a measurement problem of its own. If the claim is that earlier hidden variables are affected by later measurement settings, don't we still need a principled account of what counts as a measurement? This is a good point, but the appropriate response seems not to be to try to distinguish measurements from physical interactions in general, but to note that it is a constraint on any satisfactory development of this strategy for quantum theory that the advanced effects it envisages be products of physical interactions in general, rather than products of some special class of measurement interactions.

This point is relevant to another objection that might be brought against my suggestion. It might be argued that the approach fails to respect the core thesis of Einstein's realism, namely the idea that there is an objective physical world, whose existence and properties are independent of the choices of human observers. If our present measurements affect the prior state of what we observe, then surely the external world is not independent in Einstein's sense. If we had chosen to make a different measurement, the external world would have been different. Isn't this much like the observer dependence that Einstein found objectionable in the Copenhagen view?

I think it is best to answer this charge indirectly. First of all, I think we may assume that Einstein would not have felt that his brand of realism was threatened by the observation that human activities affect the state of the external world in all sorts of ordinary ways. The existence of trains, planes, and laser beams does not conflict with realism. The processes that produce such things are simply physical processes, albeit physical processes of rather specialized kinds. Second, it seems fair to assume that backward causation is not in itself contrary to spirit of Einsteinian realism. On the contrary, a realist of Einstein's empiricist inclinations might well think that the direction of causation is a matter to be discovered by physics.[13] But then why should the existence of backward effects of human activities be any more problematic for realism than the existence of their "forward" cousins? Provided we make it clear that in the first place the view is that certain *physical* interactions have earlier effects, and not that certain specifically *human* activities do so, the position does not seem to be one that an empiricist of Einstein's realist persuasions should object to a priori. What the view proposes, roughly speaking, is simply to extend to the case of the past a form of dependence that realists find unproblematic in the case of the future. The proposal might perhaps be objectionable for other reasons, but it does not conflict with realist intuitions of Einstein's sort.

There are some important general questions here, concerning such things as the meaningfulness of the ordinary notion of a physical state in a world which lacks the nondisturbing observational access to reality which classical physics always took for granted. Classically, it is natural to think of the state of a system as the sum of its dispositions to respond to the range of circumstances it might encounter in the future. But if the present state is allowed to depend on future circumstances, this conception seems inappropriate. I'll return briefly to some of these issues at the end of the chapter.

For the moment, however, it seems that the advanced action approach promises the usual virtues of the incompleteness interpretation of quantum mechanics favored by Schrödinger, Einstein, Bohm, and Bell himself. Unlike other versions of this view, and most if not all versions of the opposing orthodoxy, it also promises to avoid the threatened conflict between quantum mechanics and special relativity. These advantages might be expected to be offset by disadvantages elsewhere, but this does not seem to be the case. The restrictions that quantum mechanics places on measurement enable the approach to exploit a well-recognized loophole in the bilking argument, and Bell's own concerns about free will rest on very dubious grounds. (As we saw, interpreting the required failure of the independence assumption in terms of backward causation seems to sidestep these concerns very satisfactorily, at least in light of an adequate understanding of what is objectively at issue.) So the advanced action approach escapes these objections and has the independent attractions we identified in chapters 5 and 7: it restores to microphysics a temporal symmetry which we have no good reason to doubt in the first place.

Could these remarkable advantages perhaps be offset by some other argument against the hidden variable approach to quantum mechanics? In chapter 8 I described two main lines of argument against incompleteness interpretations, apart from Bell's Theorem itself. These were the so-called no hidden variable theorems and the recent GHZ argument. How do these arguments fare in light of advanced action?

ADVANCED ACTION AND THE GHZ ARGUMENT

Readers may recall that the GHZ argument is a kind of nonprobabilistic version of Bell's Theorem. It rests on a thought experiment involving three correlated particles, each potentially subject to a spin measurement in one of two directions perpendicular to its line of flight. A simple argument shows that no prior distribution of "hidden elements" responsible for the measured spins can reproduce the predictions of quantum mechanics. The experiment is not currently technically feasible, but few people doubt that the results would be as quantum mechanics predicts. At any rate, the issue of the correctness of quantum mechanics is really beside the point here. Our concern is with how quantum mechanics might be interpreted, *under the assumption that its predictions are correct.*

Like Bell's Theorem, however, the GHZ argument depends on the independence assumption. That is, it requires the assumption that the values of hidden variables do not depend on what is to happen to the particles in question in the future—in particular, on the settings of future spin measurements. In the presentation of the argument by Rob Clifton, Constantine Pagonis, and Itamar Pitowsky, for example (whose terminology I borrowed in chapter 8, in describing the Ypiarian parallel), the assumption is introduced in the following passage:

> [W]e need to assume that λ [the set of hidden variables] is compatible with all four measurement combinations. This will be so if: (a) the settings are all fixed just before the measurement events occur, so that λ lies in the backwards light-cones of the setting events, and (b) the choice of which settings to fix is determined by (pseudo-)random number generators whose causal history is sufficiently disentangled from λ.[14]

The advanced action interpretation will simply deny that the independence assumption follows from (a), however. (Proposition [b] serves to exclude the common past hypothesis, the alternative path to rejection of the independence assumption which we described above.) So the GHZ argument provides no new argument against the advanced action interpretation.

On the contrary, if anything, the GHZ argument simply adds impressive new weight to the view that the alternatives to advanced action are metaphysically unpalatable. Clifton, Pagonis, and Pitowsky make a particular point of the fact that the nonlocality they derive from the argument does not depend on assumptions about hidden variables:

> The Einstein-Podolsky-Rosen argument for the incompleteness of quantum mechanics involves two assumptions: one about locality and the other about when it is legitimate to infer the existence of an element-of-reality. Using a simple thought experiment, we argue that quantum predictions and the relativity of simultaneity require that both these assumptions fail, whether or not quantum mechanics is complete.[15]

Indeed, the argument about locality does not rely on the relativity of simultaneity, and may be regarded as showing that nonlocality is a consequence of quantum mechanics itself—once we reject the possibility that the correlations in question be explained by hidden variables. In other words, it shows that *only* a hidden variable view could escape nonlocality. True, a hidden variable solution would itself be excluded by the independence assumption, which Clifton, Pagonis, and Pitowsky take to follow from (a) and (b). However, advanced action provides reason to reject independence, even given these assumptions. In other words, the argument suggests once again that advanced action provides the only serious alternative to nonlocality.

ADVANCED ACTION AND SUPERPOSITION

Nonlocality is not the only difficulty for the hidden variable approach to quantum mechanics. As we saw in chapter 8, the most distinctive difficulties stem from a group of results known as no hidden variable theorems, which appear to show that hidden variable theories are unable to reproduce the predictions of quantum mechanics (at least in certain cases). However, the advanced action proposal provides a very direct response to these theorems. Advanced action gives us reason to abandon a fundamental presupposition of these theorems, namely that a single set of hidden variables should be able to predict the result of any *possible* next measurement on the system in question. According to the advanced action hypothesis, the hidden variables typically depend on the nature of the *actual* next measurement (or interaction generally). Had this measurement been different, the present values of the hidden variables might well have been different. Hence we should not expect the *actual* variables to reproduce the results of *merely possible* measurements.

The basic point here is such a simple one that it might seem that some important factor has been overlooked. By way of reinforcement, then, I want to connect the point with a classic philosophical discussion of the conceptual consequences of superposition in quantum mechanics—an article called "A Philosopher Looks at Quantum Mechanics," first published in 1965,[16] by the eminent Harvard philosopher Hilary Putnam. From our present perspective a particular advantage of this article is that it just predates Bell's results, and hence gives a very good sense of how the conceptual landscape of quantum mechanics appears, if issues of nonlocality are left to one side.

The relevant consequences of superposition are described by Putnam in the following passage:

> To illustrate the rather astonishing physical effects that can be obtained from the superposition of states, let us construct an idealized situation. Let S be a system consisting of a large number of atoms. Let R and T be properties of these atoms which are incompatible. Let A and B be states in which the following statements are true according to both classical and quantum mechanics:
>
> (1)   When S is in state A, 100 per cent of the atoms have property R.
>
> (2)   When S is in state B, 100 per cent of the atoms have property T –
>
> and we shall suppose that suitable experiments have been performed, and (1) and (2) found to be correct experimentally. Let us suppose there is a state C that is a "linear combination" of A and B, and that can somehow be prepared. Then classical physics will not predict anything about C (since C will, in general, not correspond to any state that is recognized by classical physics), but quantum mechanics can be used to tell us what to expect of this system. And what quantum mechanics will tell us may be very strange. For instance we might get:
>
> (3)   When S is in state C, 60 per cent of the atoms have property R,
>
> and also get:
>
> (4)   When S is in state C, 60 per cent of the atoms have property T –
>
> and these predictions might be borne out by experiment. But how can this be? The answer is that, just as it turns out to be impossible to measure *both* the position and the momentum of the same particle at the same time, so it turns out to be impossible to test *both* statement (3) *and* statement (4) experimentally in the case of the same system S. Given a system S that has

been prepared in the state C, we can perform an experiment that checks (3). But then it is physically impossible to check (4). And similarly, we can check statement (4), but then we must disturb the system in such a way that there is then no way to check statement (3).[17]

Putnam introduces what he calls the Principle of No Disturbance (ND):

> The measurement does not disturb the observable measured—i.e. the observable has almost the same value an instant before the measurement as it does at the moment the measurement is taken.

He claims that "this assumption is incompatible with quantum mechanics. Applied to statements (3) and (4) above, the incompatibility is obvious."[18]

Like many discussions of the conceptual consequences of quantum mechanics, however, Putnam's account is simply blind to the possibility of advanced action. This is revealed in particular in his formulation of the principle ND. It is quite clear that an advanced action account does take measurement to "disturb the observable measured"—and yet it does not deny what Putnam offers as an elaboration of the notion of disturbance by measurement, namely the principle that "the observable has almost the same value an instant before the measurement as it does at the moment the measurement is taken"! Putnam's explication simply ignores the kind of disturbance that the advanced action view envisages.

What makes this omission particularly significant is that the advanced action view provides a very natural explanation of the kind of statistics that Putnam describes. A simple example may help to clarify this point.

The Case of the Precognitive Cat

Suppose I live in a house with two external doors, inside each of which is a doormat. My dog spends his day sleeping on these mats, moving around occasionally. We might describe the dog's state by a pair of numbers, specifying the probability that I'll find him at the corresponding door if I choose to enter the house that way when I return home. Thus (0.5, 0.5) would be the state in which he had a 50 percent chance of being found at each of the two doors. Obviously the elements must sum to 1 (or less than 1 if we allow that the dog might not be on either of the mats.) So far we have nothing unusual, and of course we may interpret the probabilities here in terms of "classical ignorance," as quantum theorists say.

But now consider my cat, whose habits are the same, but who knows in advanced which door I will choose each day. As the cat becomes hungrier through the day, she displays an increasing tendency to place herself on the right mat (so as to greet me nicely). In early afternoon her state might be described by $(0.6, 0.6)$, say, reflecting the fact that whichever door I choose at that time, there is a 60 percent chance of finding her behind it.

Note that in the cat's case the probabilities have exactly the form of those described by Putnam: two incompatible outcomes each have a 60 percent probability of being found to be realized on measurement. The example shows that if we allow advanced action (here in the form of the cat's foreknowledge) then it is easy to give a classical interpretation of such probabilities. They are simply conditional probabilities corresponding to our "degree of ignorance" of the cat's position under each of two incompatible hypotheses. Moreover, if we treat one hypothesis as certain then we can treat the corresponding probability as unconditional, and interpret it as our "degree of ignorance" (or, better, our degree of confidence) that the cat is actually on the mat concerned.

The example also confirms that as Putnam defines his principle ND, it is not violated by an advanced action view. If I open the first door and find the cat on the mat inside it, then I know that she was on that mat an instant before I opened the door. In that sense, then, my observation has not disturbed her. In claiming that the kind of statistics that result from superposition are incompatible with ND (thus defined), then, Putnam is ignoring the possibility of advanced action.

As Putnam himself notes, quantum mechanics actually predicts not (3) and (4) but the following weaker propositions:[19]

(3′)   When S is in state C *and an R-measurement is made,* 60 per cent of the atoms have property R

and

(4′)   When S is in state C *and a T-measurement is made,* 60 per cent of the atoms have property T.

This being so, ND in Putnam's form leads to a contradiction with quantum mechanics only if we can assume that the proportion of atoms in state C which have property R at a time t is the same for atoms on which an R-measurement is made at (or just after) t as it is in general (and the same for T). However, this is effectively Bell's independence assumption, which of course the advanced action view rejects.

As is indicated by the fact that Putnam takes the independence assumption for granted, interpretations of the consequences of superposition based on its rejection have received almost no attention in the quantum mechanical literature. The Precognitive Cat provides a useful way to characterize the alternative approaches which have dominated the field. Most popular of all is the Copenhagen view, which holds in effect that it is simply an indeterminate matter which mat the cat is sleeping on, until one of the doors is opened. On this view it is not that had a given observation not been made, the system in question might have had some *different* value of the observable in question, but that it would not have had *any* value of this property.

As we have seen, the hidden variable opponents of the Copenhagen Interpretation have been equally reluctant to allow measurements to influence preexisting values of observables (or equally blind to the possibility). Putnam's analysis shows that they are therefore required to give up ND in Putnam's form. Typically this amounts to giving up continuity: hidden variable models require that measurements give rise to instantaneous and discontinuous changes in the possessed value of the property being measured. Again the cat case illustrates this possibility: it might be postulated that the act of opening a door tends to produce an instantaneous change in the position of the cat, so as to explain the nonclassical statistics of the case. It is important that this change be such that what the measurement reveals is the possessed value which results from it, rather than the value which immediately precedes it. Bohm's hidden variable theory is of this kind.

The discontinuity option preserves the principle that had a given measurement not been made, the value of the property in question *just before* the time of the measurement would have been what it is in fact. In other words, it restricts the highly nonclassical occurrences to the time of measurement itself, which perhaps explains why this approach to quantum mechanics has received very much more attention than those based on advanced action. Instantaneously affecting the present—even the distant present—is considered a great deal more plausible than affecting the past. As I have tried to make clear, however, there seems to be no sound basis for this preference. On the contrary, there seem to be excellent reasons for the opposite preference. In particular, only an advanced action approach seems able to properly reconcile nonlocality with special relativity.

Thus the Precognitive Cat illustrates that the kind of phenomena which advanced action would produce will never be free of alternative explanations. I emphasized in chapter 7 that we cannot hope to observe advanced action directly, as it were, for it depends on the possibility that our classical "leave everything as it is" notion of observation is inapplicable in certain domains. (This is what saves advanced action from the bilking argument.) So I think we cannot hope for clearer evidence of backward influence than quantum mechanics gives us, except perhaps in degree: the nonclassical correlations might be even stronger. Even with 100 percent correlations, however, the evidence would not be free of other interpretations.

So the argument for an advanced action interpretation is bound to be somewhat indirect. It is bound to rest on the interpretation's theoretical and conceptual advantages: its compatibility with special relativity, its temporal symmetry, and its ability to preserve determinateness and continuity. However, the fact that the issue has to be settled in this way does not make it illegitimate, or even exceptional by contemporary physical standards. Considerations of this kind have long played a very powerful role in physics. If this case is different, it is by virtue of the fact that it requires us to relinquish the possibility of direct observational verification, *even in principle.* Given that the rejection of this possibility is very well motivated, however, and in no sense an ad hoc move to defend the theory, it cannot be dismissed out of hand. (Certainly the advanced action proposal is challenging the presuppositions about evidence that physics has always taken for granted. However, it is doing so in a way that is integral to its argument. To refuse to consider the challenge is to dismiss the proposal without a hearing.)

Summing up, it seems that the puzzling consequences of superposition in quantum mechanics are the sort of phenomena we might have expected, if we had already given advanced action the consideration that, in hindsight, it seems to have long deserved. At least in general terms, they are the sort of phenomena to which the a priori project of chapter 7 might have led us, had we not been so much in the grip of the classical world view. Far from providing an obstacle to advanced action, then, the consequences of superposition welcome it as the key to an ancient lock.

The idea is not a new one, of course. Recall the comments of John Wheeler on the delayed choice two-slit experiment, with which I motivated our enquiry in chapter 5:

> Does this result mean that present choice influences past dynamics, in contravention of every principle of causality? Or does it mean, calculate pedantically and don't ask questions? Neither; the lesson presents itself rather as this, that the past has no existence except as it is recorded in the present.[20]

Physicists have long known that the key might fit the lock, but with very few exceptions have thought (like Wheeler himself) that it is too fantastic to be the true solution.[21] I have argued that this is a mistake, and that there are independent reasons to postulate a key of this kind. Had these reasons led us first to the key, quantum mechanics seems to be the kind of lock we ought to have expected to find.

THE ATEMPORAL VIEW

The a priori argument for advanced action rested on considerations of temporal symmetry. I argued that to appreciate these considerations it is important to address the issue from a sufficiently atemporal perspective—in other words, from the Archimedean standpoint which has motivated the book as a whole. But readers may feel that in focusing on quantum mechanics in this chapter and the last, we have lost sight of this atemporal perspective. It is difficult to talk about physics, especially in such informal and general terms, without speaking in the vernacular—that is, without taking for granted the concepts and assumptions of the ordinary view from "within" time. To finish, then, I want briefly to emphasize how natural our conclusions about quantum mechanics look from the atemporal stance. And I want ask what *kind* of mistake physics makes, to the extent that it ignores the Archimedean viewpoint.

I noted earlier that one aspect of the in-built temporal asymmetry of ordinary practice is that we take it for granted that the kinds, or categories, in terms of which we describe the world are not "forward looking." We often take account of the history of an object in describing it as being of one sort rather than another, but we never take account of its future in the same way. (We sometimes take account of present predictions concerning its future, but this is not the same thing.) This is why we find it natural to think that the class of photons which have passed a given polarizer in the past have something important in common—something in virtue of which they form a significant category, or kind, in virtue of which they are significantly different *now* from those which passed some other polarizer. However, we don't allow the same thing with respect to a future polarizer.

If we look at things from the reverse temporal perspective, this preference seems rather bizarre. If we view the same class of photons in reverse, for example, they now appear to have many different histories. Some appear to come to us via up-down polarizers, some via left-right polarizers, and so on. From this perspective it seems absurd to require that the photons be blind to these differences; and yet this is precisely what the independence assumption requires. It also seems absurd that all the photons should be in the same state as a result of an interaction that from this perspective seems to lie in their future.

Once again, the point of the reversal is to highlight the temporal asymmetry of our intuitions, and to confront us with the need to justify our double standard. We need to explain why we should prefer the intuitions we have when we look at the problem one way to those we have when we look at it in reverse. Familiarity is not enough, for it is our familiar intuitions which are under challenge.

How should we assess these principles—Wheeler's "principles of causality," for example—in light of our long discussion in the intervening chapters? There are two possible verdicts, I think, one somewhat more charitable than the other. The less charitable verdict is that the vast majority of physicists are simply mistaken, and have allowed their science to be guided by an assumption which is as groundless as the geocentric foundations of ancient cosmology. Insofar—so very, very far—as physics relies on this assumption, then, it is likely to be grossly in error.

This harsh assessment is not the only possible one, however. The more irenic verdict is that the asymmetric assumption concerned is embedded in the practice of physics, in such a way as to make it not so much *mistaken,* as seriously *partial.* In the cosmological case the comparison is with, on the one hand, the view that the geocentric cosmology is simply mistaken, on the other, the view that it may be correct so far as it goes—correct as a description of reality *from a particular standpoint,* so to speak.

The latter approach seems to me to be more useful, if only because it is sufficiently sophisticated to recognize a distinction between partiality—describing reality *from a perspective*—and other sorts of scientific error. It seems to me an attractive thought that quantum mechanics is a partial view in this sense—partial not in the sense that it is simply incomplete, but in the sense that it has to be understood as a view of the world from a particular standpoint. Roughly speaking, we might say that quantum mechanics represents an idealized codification of all the information about a system available to an observer who is herself embedded in time, in virtue of interactions between the system and the world in that observer's past. Thus it is a complete description from that standpoint, though an incomplete one to God. (On this reading, the mistake lies in a failure to challenge the implicit classical assumption that the perspective we have as creatures in time is not significantly partial or incomplete, at least in principle. In principle, classical observation makes us gods.)

What would an atemporal physics be like? To answer this question we need to examine the ways in which the concepts ordinarily applied in physics reflect our temporal standpoint. We made a start on this in chapters 6 and 7, in effect, in trying to dissect out the conventional components in our ordinary notions of counterfactual dependence and causation. Much more needs to be done. The conceptual apparatus of physics seems to be loaded with the asymmetric temporality of the ordinary world view. Notions such as *degree of freedom, potential,* and even *disposition* itself, for example, seem to embody the conception of an open future, for which present systems are variously prepared. (We don't speak of bodies being presently disposed to behave in certain way if appropriate conditions have eventuated in the past.) As I noted earlier, the very notion of a *state* often seems to be forward-looking in this way—a fact which partly explains why the independence assumption has so rarely been challenged, I think. The assumption seems to be built in to the conceptual machinery, in a way in which we have only been able to ignore at the cost of some awkwardness. We normally think of measurements as activating preexisting dispositions in things to affect apparatus in certain ways. This too has had to go to accommodate advanced action, and it is far from clear how best to characterize what must replace it.

These problems do not give us reason to relinquish the insights of the advanced action approach, of course. On the contrary, to see the problems is to see the perspectival character of the ordinary viewpoint, and hence to grant one of the main claims on which the advanced action view relies (namely that ordinary ways of thinking embody little-recognized temporal asymmetries). Once we reach this point, then, there is no going back. We are committed to the program of building an atemporal Archimedean physics, of developing the conceptual machinery such a physics requires. The aim of this chapter has been to show that the effort is likely to be worthwhile. Leaving aside the more rarefied philosophical attractions of this atemporal view, it promises profound insights into the some of the deepest problems of contemporary physics.

10

$\longleftarrow\longrightarrow$

Overview

Aт the beginning of the book I described two opposing viewpoints in
the philosophy of time. One view holds that the present moment and
the flow of time are objective features of reality. The other view disagrees,
treating the apparent objectivity of both these things as a kind of artifact
of the particular perspective that we humans have on time. According to
the latter view what is objective is the four-dimensional "block universe," of
which time is simply a part. In chapter 1, I outlined some of the attractions
of the block universe view. Since then, the project of the book has been to
explore its consequences in physics, in two main respects: first, in connection with the attempt to understand various puzzling temporal asymmetries
in physics; and second, by way of its bearing on various time-asymmetric
presuppositions, which turn out to play a crucial role in standard ways of
thinking about quantum mechanics.

In particular, I have been trying to correct a variety of common mistakes
and misconceptions about time in contemporary physics—mistakes and misconceptions whose origins lie in the distorting influence of our own ordinary
temporal perspective, and especially of the time asymmetry of that perspective. One important aspect of this problem is a matter of sorting out how
much of the temporal asymmetry we think we see in the world is objective,
and how much is simply a by-product of our own asymmetry. I have urged
that in order to clarify these issues, and to avoid these mistakes, we need to
learn to set aside some very deeply ingrained habits of thought. We need to
familiarize ourselves with an atemporal perspective—an Archimedean "view
from nowhen."

The physical and philosophical concerns of the book have thus been very
closely intertwined. The book's conclusions have emerged at a variety of
levels, in a variety of voices. Some were substantial proposals concerning

contemporary problems in physics or philosophy, others were prescriptions for the proper conduct of these disciplines from the Archimedean standpoint, and so on. In order to help readers to put the whole thing in perspective, I have listed below, by chapter, the main conclusions of the book.

In this book, especially, it would be out of character if the overview looked only in one direction. I finish, therefore, with a few pointers to future work—to the kinds of issues that look important in physics and philosophy, in light of these conclusions.

MAIN CONCLUSIONS OF THE BOOK

CHAPTER 2. *The Lessons of the Second Law*

• What needs to be explained is the low-entropy past, not the high entropy future—why entropy goes down toward the past, not why it goes up toward the future.

• To a significant extent, then, the *H*-theorem and its descendants address a pseudo-problem.

• The traditional criticism of the *H*-theorem—viz., that it assumes temporal asymmetry in disguised form—turns out to be well motivated but misdirected. The important issue is not whether we are entitled to assume the *stoßzahlansatz* (or PI^3, the Principle of the Independence of Incoming Influences) toward the future, but why these independence principles do not hold toward the past.

• We need to guard against the double standard fallacy—that of accepting arguments with respect to one temporal direction which we wouldn't accept with respect to the other.

• The most useful technique for avoiding these fallacies involves imagined time reversal. If an apparently acceptable argument looks counterintuitive when we imagine time reversed, it is a good indication that a double standard is in play. In effect, this simple technique provides temporal creatures such as ourselves with a reliable and readily accessible guide to the standards that would apply from a genuinely atemporal perspective.

CHAPTER 3. *New Light on the Arrow of Radiation*

• The issue concerning the asymmetry of radiation is sometimes misrepresented. Correctly understood, it is that as to why there are large coherent sources in the past but not (apparently) in the future.

• A proper understanding of the problem of temporal asymmetry in thermodynamics shows that a common argument which claims to derive this asymmetry of radiation from the thermodynamic behavior of matter (e.g., the edges of ponds) is fallacious, for it needs to assume the absence of the very boundary conditions—viz., coherent sources of advanced radiation—that it seeks to exclude.

• This fallacy is even more serious in the Wheeler-Feynman Absorber Theory, which explicitly assumes that there really is advanced radiation, although we don't see it.

• The issue of the asymmetry of radiation thus turns out to be parallel to (rather than *reducible to*) that raised by thermodynamics, in the sense that it too directs us to the existence of highly ordered conditions in the past.

• This diagnosis of the nature of the asymmetry of radiation is confirmed by our reinterpreted version of the Wheeler-Feynman theory, which shows that radiation can be considered to be symmetric at the micro level.

• The argument for the proposed reinterpretation reveals other flaws in the standard version of the Wheeler-Feynman theory.

CHAPTER 4. *Arrows and Errors in Contemporary Cosmology*

• The asymmetries of thermodynamics and radiation appear to depend on the fact that the universe had a particular character early in its history: its matter was very evenly distributed, which is a very ordered condition for a system in which gravity is the dominant force.

• Contemporary cosmologists continue to underestimate the difficulty of explaining this condition of the early universe without showing that the universe must be in the same condition at its other temporal extremity (which would imply that the familiar asymmetries would reverse as the universe recollapsed). Blindness to this difficulty—the *basic dilemma,* as I called it—stems from double standard fallacies.

• Many arguments against the symmetric collapse model also involve double standard fallacies, particularly in relying on statistical reasoning which would equally exclude a low-entropy big bang.

• There are important questions concerning the consistency and observability of a time-reversing collapse which—because it has been rejected on spurious grounds—have not been properly addressed by physics.

• Although in many ways further advanced than it was in the late nineteenth century, the contemporary discussion of temporal asymmetry in physics is still plagued by some of the same kinds of mistakes.

CHAPTER 5. *Innocence and Symmetry in Microphysics*

• It is important to distinguish two forms of PI³: the macroscopic case, associated with the fact that the universe has a low-entropy past, and a microscopic case, almost universally taken for granted in physics. The microscopic case embodies the intuitively plausible principle of μInnocence: interacting systems are uncorrelated before they first interact.

• Unlike its macroscopic cousin, the acceptance of μInnocence does not rest on observational grounds. As it currently operates in physics, it is an independent asymmetric principle, in conflict with the assumed T-symmetry of (almost all) the underlying laws of microphysics.

• Hence there is a deep and almost unrecognized conflict in contemporary physics. If we are to retain T-symmetry, we should abandon μInnocence.

• Quantum mechanics suggests that there might be good independent reasons for abandoning μInnocence. μInnocence turns out to be a presupposition of the main arguments for thinking that there is something especially puzzling about quantum mechanics. In other words, quantum mechanics seems to offer empirical confirmation that μInnocence fails.

• The failure of μInnocence seems to open the way for a kind of backward causation. However, well-recognized features of quantum mechanics seem to block the paradoxes to which backward causation is often thought to lead. But the suggestion raises wider issues about the asymmetry of causation itself, which need to be addressed in their own terms, before the proposal concerning μInnocence can be evaluated properly.

CHAPTER 6. *In Search of the Third Arrow*

• Although the asymmetry of causation is often said by physicists to be of no relevance to contemporary physics, it continues to exert a great influence on the practice of physics. Hence its interest is not merely philosophical: it needs to be understood, so that this influence may be assessed.

• The most popular philosophical approach to the asymmetry of causation is the third arrow strategy, which seeks to analyze causal asymmetry in terms of a de facto physical asymmetry. However, it turns out that the available candidates are not appropriately distributed in the world. In particular, they fail at the micro level.

• This point is often obscured by fallacies similar to those which plague attempts to account for the physical temporal asymmetries: double standards and buck-passing, for example.

• The most plausible solution is the anthropocentric one: the asymmetry of causation is a projection of our own temporal asymmetry as agents in the world.

CHAPTER 7. *Convention Objectified and the Past Unlocked*

• The diagnosis of the previous chapter finds attractive expression in terms of the conventional asymmetry of counterfactual conditionals. However, the conventionalist view seems to make the asymmetry of dependence—the fact that the future depends on the past, but not vice versa—insufficiently objective, in two senses: it seems too weak, in making the asymmetry conventional, and too strong, in ruling out backward causation by fiat.

• The conventionalist view meets the first point by noting that the convention is not a matter of choice, and thereby explaining its apparent objectivity.

• The conventionalist view meets the second point by showing that there is a loophole which allows backward dependence, in circumstances in which an agent's access to past events is limited in certain ways.

• The admission of backward dependence requires an appropriate disambiguation of the relevant convention governing our use of counterfactuals. The disambiguation in question is a matter of linguistic choice, but it is an objective matter whether the world is such as to require us to make this choice.

• Hence there is an objective possibility concerning the way in which the microworld is structured, which has been all but obscured by our familiar intuitions concerning causation, μInnocence, and the like. As in chapter 5, moreover, it turns out that there is a strong symmetry argument in favor of the hypothesis that the microworld actually has a structure of this kind.

• Temporal symmetry alone might thus have led us to expect a kind of backward causation, or *advanced action,* in microphysics.

CHAPTER 8. *The Puzzle of Contemporary Quantum Theory*

This chapter presented a broad overview of the conceptual issues concerning the interpretation of quantum mechanics, emphasizing the central role of the issue as to whether quantum mechanics is complete. In setting out the difficulties faced by the competing approaches to this issue, my exposition mainly followed conventional lines, but made a few distinctive claims:

• I argued that hidden variable approaches are in a stronger position than is usually recognized. Given that all conventional views admit nonlocality, it is not a decisive objection to hidden variable views that they too are required to do so. In terms of the conventional debate—the debate which ignores advanced action—then, the contextualist approach remains underexplored.

• I noted that no collapse views face a difficulty concerning the meaning of probability in quantum mechanics which is even more severe than has previously been recognized, even by philosophical critics.

CHAPTER 9. *The Case for Advanced Action*

• Bell's Theorem depends on the independence assumption, which might be relaxed in two ways: dependence may be secured either in the past, via a common cause, or in the future, via the kind of advanced action whose formal possibility we identified in chapter 7. If successful, either of these strategies would enable quantum mechanics to avoid nonlocality.

• The common cause strategy seems initially the more attractive strategy in light of our ordinary causal intuitions, but calls for an implausible substructure underlying ordinary physical processes.

• The advanced action is elegant and economical is comparison, and has the symmetry advantage noted in chapter 6. Quantum mechanics supplies the restrictions on classical observability that the argument of chapter 6 led us to expect.

• The benefits of the advanced action proposal are not confined to Bell's Theorem; the proposal also undercuts the non-EPR no hidden variable theorems, and the new GHZ argument for nonlocality.

• Quantum mechanics might be interpreted as providing a complete description *from a limited or partial perspective:* a complete view of the world *as accessible from the temporal standpoint we normally occupy.* This is compatible with the claim that it is an incomplete description of what would be seen from the Archimedean standpoint.

• This suggestion raises important issues concerning the extent to which the ordinary conceptual framework of physics depends on the temporal viewpoint, for example, in its use of concepts such as *degree of freedom* and *potential,* and methods such as statistical reasoning. In this respect the proper form of an atemporal "physics from nowhen" is a issue left open by this book.

DIRECTIONS FOR FURTHER WORK

What sorts of projects look important in the light of these conclusions? There is work for both physicists and philosophers, I think.

In physics

• Exploration of models incorporating advanced action, especially in quantum mechanics.

• Exploration of the consistency and possible empirical consequences of symmetric time-reversing cosmologies, and more generally of the issue of the observability of phenomena constrained by future low-entropy boundary conditions.

• The project of explaining the low-entropy big bang, with the basic dilemma clearly in view.

In philosophy

• The issue of the proper conceptual framework for an atemporal physics. How much of the conceptual machinery of conventional physics depends on our familiar temporal perspective?

• Similar issues in metaphysics more generally. I have argued that causation and physical dependence are importantly anthropocentric notions, whose temporal asymmetry reflects the contingencies of our own temporal stance. But what would a properly atemporal metaphysics be like?

WHY IT MATTERS

In what sense do these issues matter? Why shouldn't we ignore the view from nowhen, and go on in physics, philosophy, and ordinary life just as we always have? After all, we cannot actually step outside time, in the way in which we can climb a tree to alter our viewpoint. Isn't it better to be satisfied with the viewpoint we have?

We cannot step outside time, but we can try to understand how the way in which we are situated within time comes to be reflected in the ways in which we talk and think and conceptualize the world around us. What we stand to gain is a deeper understanding of ourselves, and hence—by subtraction, as it were—a deeper understanding of what is external to us. This is a reflective kind of knowledge: we reflect on the nature of the standpoint from within, and thereby gain some sense—albeit, ultimately, a sense-from-within—of what it would be like from without.

If the reflexivity were vicious the project would be self-defeating, but is it vicious? Our understanding seems to be enhanced, not overturned. The issue here is an old one: science has long stood proxy in this way for creatures—ourselves—whose own epistemological connections with the world are tenuous, patchy, contingent, and parochial. With each advance comes a new picture of how the world would look from nowhere, and a new appreciation of the limits of our own standpoint. At each stage there is a temptation to think that our standpoint is devalued, but this seems to be a mistake. If we had a choice of standpoints we might choose a different one, but to be swayed by this would be like wanting to be someone else.[1] Because our standpoint is not a matter of choice—no more so than it is a matter of choice who we are—it cannot coherently be undermined in this way.

The campaign for a view from nowhen is a campaign for self-improvement, then, and not a misguided attempt to do the impossible, to become something that we can never be. It promises only to enhance *our* understanding of ourselves and our world, and not to make us gods.

Notes

CHAPTER 1. *The View from Nowhen*

1. *Confessions,* Book XI.14. Augustine (1912), p. 239.

2. *The View from Nowhere* is Nagel's 1986 book.

3. See Williams (1951), Smart (1955, 1963), Grünbaum (1973), and Mellor (1981), for example.

4. A good recent discussion of these issues, with some references to earlier work, is that of Horwich (1987), pp. 33–36.

5. See Davies (1974), p. 176, (1995), pp. 208–13, and Sachs (1987).

6. As we shall see, a good case can be made for the view that the basic puzzle lies in cosmology—in particular, in the question why the universe had very low entropy, early in its history. It has been suggested that the asymmetry of the neutral kaon can help with other cosmological puzzles, such as why there seems to be more matter than antimatter in the universe—see Davies (1995), p. 213, for example—but the issue of a possible connection with the low-entropy start remains unclear.

7. *Il Saggiotore,* from a passage quoted by E. A. Burtt (1954), p. 85. Similar thoughts had been voiced much earlier. Democritus is said to have held that reality is simply a matter of "atoms and the void," and that all else is "by convention." But as Burtt emphasizes, Galileo's version of the doctrine introduces a new distinction between the objective realm of concern to mathematical science, on the one hand, and the subjective realm of the secondary qualities, on the other: *"In the course of translating this distinction of primary and secondary into terms suited to the new mathematical interpretation of nature, we have the first stage in the reading of man quite out of the real and primary realm"* (Burtt 1954, p. 89, italics in the original).

8. This is envisaged in a number of contemporary cosmological models.

CHAPTER 2. *The Lessons of the Second Law*

1. Brush (1976), p. 551. As will be apparent to readers who know Brush's work on the history of thermodynamics, the historical material in this chapter is much indebted to his definitive treatment of the subject.

2. Newton (1952), p. 398, quoted in Brush (1976), p. 545.

3. In the absence of which the irreversibility might plausibly be attributed to unknown *asymmetric* forces.

4. Hunter (1788), p. 53, quoted in Brush (1976), p. 554.

5. Thomson (1852).

6. Clausius (1865). This is the paper in which the term "entropy" is first introduced, though as Brush (1976), p. 576 points out, Clausius was using the concept, if not the term itself, by 1854.

7. Boltzmann (1877), p. 193, in the translation in Brush (1966).

8. Culverwell (1890a, 1890b, 1894).

9. Burbury (1894, 1895).

10. Burbury (1894), p. 78.

11. See Boltzmann (1895b).

12. Zermelo (1896). For a detailed account of the recurrence objections, see Brush (1976), chapter 14.7.

13. See Boltzmann (1895a), p. 415, and (1964), p. 446.

14. An appreciation of this point seems to be rather rare in the literature. However, as I am grateful to Phillip Hart for pointing out to me, it is made forcefully in von Weizsäcker (1939), a paper which appears in English translation as §II.2 in von Weizsäcker (1980). Von Weizsäcker notes that "improbable states can count as documents [i.e., records of the past] only if we presuppose that still less probable states preceded them." He concludes that "the most probable situation by far would be that the present moment represents the entropy minimum, while the past, which we infer from the available documents, is an illusion" (1980, pp. 144–45).

15. Von Weizsäcker (1980) attributes this point to Bronstein and Landau (1933).

16. One of the few writers who seems to have properly appreciated this point is Roger Penrose, who says "The high-entropy states are, in a sense, the 'natural' states, which do not need further explanation" (1989, p. 317).

17. I am grateful to Professor Philip Pettit for permission to use this joke.

18. However, there is a different way in which temporal asymmetry might arise from a symmetric physics, as we shall see in chapter 4.

19. The notion of a branch system is due to Hans Reichenbach (1956), who uses it to address what he sees as the difficulties raised by the reversibility objections. Reichenbach endorses a viewpoint very much like Boltzmann's own time-symmetric view. Grünbaum (1973) also employs this notion, but makes stronger use than Reichenbach of what is in effect PI^3. His view is hence more vulnerable to the kind of objections which follow.

20. Sklar (1992), p. 145.

CHAPTER 3. *New Light on the Arrow of Radiation*

1. Zeh (1992), p. 12.

2. Einstein and Ritz (1909), translation from Beck (1989), p. 376.

3. Zeh (1992), p. 14.

4. Popper (1956a).

5. Popper later noted (1956b) that he was unaware of the Ritz-Einstein debate at the time he wrote his (1956a) letter to *Nature*.

6. My discussion of the Absorber Theory will be restricted to the framework of classical electrodynamics. A number of writers have discussed the Absorber Theory in a quantum framework—see, e.g., Cramer (1980, 1883, 1986), Davies (1970, 1971, 1972), and Hoyle and Narlikar (1974). The question whether the present argument may be expected to extend to the quantum case is one of several issues discussed briefly at the end of the chapter.

7. Davies (1974), p. 119.

8. Davies (1974), p. 119; Zeh (1992), p. 19.

9. Zeh (1992), p. 13.

10. This is very much the same as Popper's conclusion about the relations between thermodynamics, radiation, and initial conditions. My objection to Popper turned on the argument he appears to endorse concerning the impossibility of the occurrence of the reverse of these asymmetric phenomena, not on his view of the relation of radiation to thermodynamics.

11. Zeh (1989), p. 13, and in correspondence.

12. See Zeh (1992), pp. 3, 13. This is the suggestion that I want to defend below. A particularly clear statement of the view is to be found in Stephenson (1978). However, to my knowledge other writers have not pointed out that the Wheeler-Feynman theory may be reinterpreted so as to support this position, by guaranteeing the necessary consistency between emitters and absorbers; nor that this view leads to a particularly simple reduction of the problem of the apparent asymmetry of radiation to that of cosmological boundary conditions.

13. Davies (1977), p. 181.

14. Davies (1974), p. 112.

15. There is an interesting historical account of their collaboration on this theory in James Gleick's recent biography of Feynman; see Gleick (1992), pp. 110–26.

16. See Davies (1977), p. 178, for example.

17. For a more detailed treatment of the theory see Wheeler and Feynman's original (1945) paper, or Paul Davies' account in (1974), chapter 5.

18. Davies (1974), p. 144.

19. Wheeler and Feynman (1945), p. 170.

20. What happens to the component of the "response" from the absorber in the spacetime region prior to the original emission at *i?* In the Wheeler-Feynman version of the argument this component was mutually canceled out by the one-half advanced wave from *i*, but in my picture the latter is not available. Isn't there a problem here? No, for look at the problem from the reverse temporal perspective: the wave from the absorber, which now appears as an incoming concentric wave centered on *i*, now appears to be *fully absorbed* at *i*, so that there is no surviving component in what now appears to be the region after *i*. Remember that it is really the same wave that from the normal perspective seems to originate at *i*. We don't think it puzzling that that wave doesn't exist *before* the source event at *i*.

21. See Burman (1970, 1971), Cramer (1983), Hogarth (1962), Hoyle and Narlikar (1974), and the contributions of Hogarth, Hoyle, and Narlikar in Gold (1967), for example.

22. Burman (1972, 1975) and Narlikar (1962) discuss the case of neutrinos, while Csonka (1969) aims to include both massless and massive particles.

23. Csonka (1969), p. 1267.

24. It has sometimes been suggested that the claim just made depends on a classical framework, and that the move to a quantum framework makes a significant difference in our ability to pose the problem of radiative asymmetry—in particular, in our ability to draw an unambiguous distinction between advanced and retarded radiation. Thus Cramer (1986, p. 661, n. 15) appears to reject the classical identification of the emission of a wave "traveling backwards in time" with the absorption of a wave "traveling forwards in time" on the grounds that in the quantum formulation, advanced and retarded waves have different time-dependent phases, namely $\exp(i\omega t)$ and $\exp(-i\omega t)$, respectively. However, this difference simply reflects the fact that the phase is specified with respect to the time of absorption in one case and with respect to the time of emission in the other. (An analogous phase difference between advanced and retarded electromagnetic potentials and associated antenna currents is a feature of the classical theory of emitters and absorbers; see Stephenson 1978, p. 923, for example.) It corresponds to the real difference between (1) an incoming wave absorbed at a point at a time and (2) an outgoing wave emitted at that point at that time; not to the illusory distinction between (3) a wave which is "really" incoming in the usual temporal sense and (4) a different wave which occupies the same spacetime cone but is "really" outgoing in the reverse temporal sense. Thus the quantum treatment does not appear to introduce any novel element in this respect.

25. Cramer (1980, 1986).

Chapter 4. *Arrows and Errors in Contemporary Cosmology*

1. There is an excellent account of these arguments in Penrose (1989), ch. 7.

2. This work is described in Davies (1995), ch. 6, for example.

3. See Gold (1962), for example.

4. Davies (1977), pp. 193–94.

5. See Penrose (1989), ch. 7.

6. Penrose himself also makes the point in this way—see Penrose (1989), p. 339, for example. However, we shall see that he sometimes fails to appreciate the logically equivalent contrapositive point, which is that if we take statistical reasoning to be inappropriate toward the past, we should not assume that it is appropriate with respect to the future.

7. For a general introduction to the inflationary model see Linde (1987).

8. Davies (1983), p. 398

9. This is close to a point raised by Page (1983). Page objects that in arguing statistically with respect to behavior during the inflationary phase, Davies is assuming the very time asymmetry which needs to be explained—assuming, in effect, that entropy increases rather than decreases. However, I think Davies might reply to this that statistical reasoning is more fundamental than the thermodynamic asymmetry itself, and that it is perfectly acceptable in the absence of constraining boundary

conditions. There would be two possible responses at this point. One might argue (as Page does) that initial conditions have to be special to give rise to inflation in the first place, and hence that Davies' imagined initial conditions are in fact far from arbitrary. Or more directly, one might argue as I have, that if there is no boundary constraint at the time of transition from inflationary phase to classical big bang, then we are equally entitled to argue from the other direction, with the conclusion that the universe is inhomogeneous at this stage.

10. Davies (1983), p. 399.

11. All the same, it might seem that there is an unresolved puzzle here: as we approach the transition between an inflationary phase and the classical phase from one side, most paths through phase space seem to imply a smooth state at the transition. As we approach it from the other side most paths through phase space appear to imply a very nonsmooth state. How can these facts can be compatible with one another? I take it that the answer is that the existence of the inflationary phase is in fact a very strong boundary constraint, invalidating the usual statistical reasoning from the "future" side of the transition.

12. Hawking (1988).

13. Hawking (1988), p. 148.

14. Hawking (1988), p. 150.

15. Hawking (1988), p. 150.

16. Price (1989).

17. Zeh (1989, 1992).

18. Halliwell (1994), pp. 382–83, Davies (1995), p. 231. Davies suggests that the loophole depends on the many worlds view of quantum mechanics, but I think this is a mistake: as my simple analogies illustrate, the required consequence could be a feature of quite ordinary theories.

19. See Hawking (1994), pp. 347–48, for example.

20. This loophole may be smaller than it looks. Hawking's no boundary condition would not provide an interesting explanation of temporal asymmetry if it simply operated like the assumption that all allowable models of the universe display the required asymmetry. This would amount to putting the asymmetry in "by hand" (as physicists say), to *stipulating* what we wanted to *explain*. If the no boundary condition is to exploit this loophole, in other words, it must *imply* this asymmetry, while being sufficiently removed from it so as not to seem ad hoc.

21. Hawking (1994).

22. Hawking (1994), p. 350.

23. Hawking (1994), p. 351.

24. Hawking (1994), p. 354; at this point Hawking acknowledges the work of Lyons (1992).

25. Hawking (1994), p. 355.

26. Hawking (1994), p. 355.

27. Halliwell and Hawking (1985), p. 1777.

28. Halliwell and Hawking (1985), p. 1778.

29. Hawking (1985), p. 2494.

30. See Hawking (1994), pp. 347–48, for example.

31. See note 20.

32. Hawking (1985), p. 2490.

33. See Penrose (1979), and particularly Penrose (1989), ch. 7.

34. The issue would be different if Penrose could exploit the tiny asymmetry which seems to be already present in physics, viz., the case of the neutral kaon, which I mentioned in chapter 1. Davies (1995), p. 218, reports that Penrose thinks there might be a connection between the two cases; if so, then Penrose's argument would certainly be on stronger grounds than I suggest in the text.

35. See Penrose (1979), pp. 597–98, and Hawking (1985), p. 2491, for example.

36. The problem does not arise if all possible worlds are equally real, and none is "actual," except from its own point of view, as in David Lewis's (1986b) theory. This kind of view does not avoid the ontological profligacy mentioned in the text, however.

37. Penrose (1979), p. 634.

38. See Davies (1974), p. 96, for an argument of this kind.

39. Davies (1977), p. 196.

40. Davies (1977), pp. 195–96.

41. Davies (1977), p. 193.

42. Davies (1974), p. 199.

43. Penrose (1979), pp. 598–99; a similar argument appears in Penrose (1989), pp. 334–35.

44. Hawking (1985), p. 2490.

45. Penrose (private communication, 1991).

46. Penrose (private communication, 1991).

47. The same goes for an older argument against the Gold view, which points out that "in the normal course of events" radiation won't reconverge on stars. In the normal course of events it wouldn't do so in the reverse direction either, but something seems to override the statistical constraint. See Davies (1974), p. 96, for this argument (based on earlier work by Martin Rees). Davies also has another radiation-based argument against the Gold universe: "Any photons that get across the switch-over unabsorbed will find when they encounter matter that the prevailing thermodynamic processes are such as to produce *anti-damping*. ... If a light wave were to encounter the surface of a metallic object, it would *multiply* in energy exponentially instead of diminish. ... Consistency problems of this sort are bound to arise when oppositely directed regions of the universe are causally coupled together." (Davies 1974, p. 194) Apparently, however, Davies no longer regards this as a powerful argument against the Gold view, for in Davies and Twamley (1993) he and Jason Twamley canvas other ways in which radiation might behave in a time-reversing cosmos. I turn to the most significant of these arguments in the next section.

48. I am indebted here to Steve Savitt's class at UBC in 1995, who helped me to get this point straight.

49. Gell-Mann and Hartle (1994), pp. 326–27.

50. Davies and Twamley (1993).

51. This experiment should not be confused with a well-known test of some cosmological implications of the Wheeler-Feynman Absorber Theory of Radiation. One of the predictions of the Absorber Theory is that a transmitter will not radiate at full strength in directions in which the future universe is transparent to radiation. This was interpreted to mean that by looking for directions in which transmitters would not transmit at full strength, one could look for "holes" in the future universe—regions in which the universe is transparent to radiation. Partridge (1973) performed a version of this experiment, but found no such "holes." (For an informal description of Partridge's experiment, see Davies 1995, p. 203.) But the kind of observation I have described does not depend on the Absorber Theory, which seems to me misconceived, as I explained in chapter 2. It also predicts a quite different result, namely *increased* rather than *decreased* radiation in certain directions. And it isn't looking for the holes in the future universe, but the exact opposite: the stars and galaxies associated with what we see as the contracting phase of a Gold universe.

52. This conclusion could be avoided by showing that gravitational collapse does not naturally lead to a high-entropy singularity at all—in other words, by finding within one's theory of gravity an argument to the effect that entropy naturally *decreases* in gravitational collapse. Sikkima and Israel (1991) have claimed to show that a low-entropy state may indeed be the natural result of gravitational collapse. However, they do not see the argument as supporting the time-symmetric Gold view. For one thing, they say that in a cyclical universe entropy will increase from cycle to cycle. So the old puzzle would reemerge at this level: How can such an overall asymmetry be derived from symmetric assumptions?

CHAPTER 5. *Innocence and Symmetry in Microphysics*

1. Penrose and Percival (1962), Davies (1974), p. 119. As their terminology suggests, Penrose and Percival themselves take the view that conditional independence has the status of a law of physics, and therefore reject T-symmetry. As I have explained, however, this aspect of their view has not caught on: most physicists regard the asymmetry of conditional independence as a matter of boundary conditions.

2. For an accessible introduction to these ideas see Horwich (1987), for example.

3. Bohm's theory is first described in Bohm (1952). For an accessible recent account of the theory and its advantages, see Albert (1994). Other discussions include those of Putnam (1979), pp. 140, 145, Jammer (1974), ch. 7, Bell (1981), pp. 57–58, and Albert (1992), ch. 7.

CHAPTER 6. *In Search of the Third Arrow*

1. The qualification here is intended to take account of the possibility of "backward" causation or dependence. This possibility is controversial, of course, but in order not to rule it out of court at this stage, it seems better not to insist that the asymmetries we are dealing with are universal rather than merely predominant in nature. We shall return to the possibility of backward causation in chapter 7.

2. Hawking (1994), p. 346.

3. Wheeler (1978), p. 41, my italics.

4. Arguments of this kind may be found, for example, in Horwich (1987), p. 8, Hausman (1986), p. 143 and Papineau (1985), pp. 273–74. See also Lewis (1986a), pp. 40–41.

5. Except where this correlation is itself explicable in terms of an earlier divergence from a common center, of course.

6. What counts as a "suitable central event"? The intuitive idea is that it be an event which is correlated with each of the other two events. At this stage the third arrow strategist doesn't want to fill out this idea in causal terms, however.

7. See Reichenbach (1956), p. 157.

8. See Reichenbach (1956), pp. 158–63.

9. Arntzenius (1990), p. 95, my italics.

10. Lewis (1986a), p. 32.

11. Lewis (1986a), p. 50.

12. Lewis (1986a), p. 50.

13. *Strict* overdetermination is not a matter of degree, it should be noted. As Arntzenius (1990) points out, each event is in general determined by nothing less than a complete later time slice, and this fact is entirely symmetric.

14. A common response to this argument is to suggest that the third arrow strategy could simply take the macroscopic fork asymmetry to provide a kind of temporal "signpost," with respect to which even microscopic events could be classified as cause and effect. By definition, then, the effect is whichever member of a pair of suitably related events lies in the temporal direction in which macroscopic forks are "open." This is close to the approach suggested by Reichenbach himself (1956, p. 127), but it is simply another form of Humean conventionalism. Indeed, it is even more arbitrary than Hume's original, in a sense, for it simply aligns our use of the notions of cause and effect to what, as it stands, is simply a de facto correlate of the past-future distinction. (It is a bit like defining "Act I" as the part of the play which is performed closest to lunch time on the day of performance.) This could be turned to the suggestion's advantage, however, if it explained the point of the convention concerned in terms of the de facto asymmetry in question. But the result would be a better-motivated version of conventionalism—such as I endorse below, in effect—and not a version of the third arrow strategy.

15. Russell (1963), p. 132.

16. Russell (1963), p. 147, italics in the original.

17. See Tooley (1987) for an account of this kind.

18. It has often been noted that David Lewis's account of causation is vulnerable to this sort of challenge, by virtue of his reliance on the notion of real possible worlds, causally distinct from the actual world in which we live. See Blackburn (1993), p. 73, for example, and van Fraassen (1989). My criticism of Lewis's theory above is independent of this issue, however.

19. See, for example, Wright (1993), Johnston (1989, 1993), Pettit (1991), and Price (1991e). One focus of these discussions has been the question as to how best

to represent the subjectivity apparently characteristic of the traditional secondary qualities; another, the issue as to how much of our discourse actually exhibits this or related forms of subjectivity.

20. Even if it is an option with respect to the traditional secondary qualities, it may not be with respect to other concepts, which—though displaying an analogous kind of subjectivity—play a more significant role in science. In my view causation is one such concept, and probability another. For a brief discussion of the subjectivity of the latter concept, see the first part of Price (1991a).

21. Ramsey (1978), p. 146. Other advocates of the agency view include Colling-wood (1940), Gasking (1955), von Wright (1975), and Price (1991b, 1992a, 1992b, 1993).

22. See Menzies and Price (1993) for this line of argument.

23. Horwich (1987), pp. 201–2.

24. Perhaps the surprising thing is that this conclusion does not already seem intuitively plausible. I think it is an indication of the extent to which twentieth-century physics has muddied the waters concerning both causation and temporal asymmetry that (unlike many physicists) contemporary philosophers don't find it natural to deny that such an asymmetric relation as causation is an intrinsic feature of the physical world.

CHAPTER 7. *Convention Objectified and the Past Unlocked*

1. See Jackson (1977), for example.

2. Lewis (1986a), pp. 40–41.

3. To avoid later confusion, let me emphasize that there is another aspect to the intuitive asymmetry embodied in the photon case, beyond the asymmetry captured by the contrast between 7·1 and 7·2. In a sense, then, I agree that the convention-alist approach does not account for all the *apparent* objective asymmetry of such a case. However, the extra asymmetry is that of μInnocence, which I think we are wrong to expect microphysics to respect. In order to establish the vulnerability of μInnocence, we need to show that there is no better-grounded objective asymmetry, say, of counterfactual dependence, behind which it might shelter.

4. Some of this phenomenology might perhaps be explained in evolutionary terms, as suggested by Horwich (1987), p. 187, for example. At any rate, if it didn't seem like this, then it wouldn't be deliberation.

5. There is no necessity in this, of course. We could, and in practice certainly do, use counterfactuals for other purposes. The claim is simply that in those contexts in which the symmetry of dependence seems vivid to us—contexts such as that of 7·1 and 7·2 above—the use concerned is one which does depend on this connection with deliberation.

6. Note that what we are interested in showing is that it is an empirical possi-bility that the world might contain what *we* would describe as backward causation; not merely that there might be differently oriented creatures who would see it as containing what we would have to describe—to the extent that we could describe

it—as backward causation. The issue is whether the conventionalist proposal makes sense of the idea that even *from our own perspective* it is a posteriori that we can't affect the past.

7. I am simplifying here, of course. For one thing it is clear that even given the hypothesis of time travel, we are never actually justified in expecting the experiment to yield contradictory results, for logic alone rules that out. A number of authors have made this the basis of a defense of the possibility of time travel against the bilking argument. See Horwich (1975), Lewis (1976) and Thom (1974), for example. This issue is not directly relevant to our present concerns, which exploit a much larger loophole in the bilking argument. In passing, however, let me record my view—similar to that of Horwich (1987), ch. 7—that the bilking argument survives the former challenge. Roughly speaking, it shows us that the hypothesis of time travel can be made to imply propositions of arbitrarily low probability. This is not a classical reductio, but it is as close as science ever gets.

8. See Dummett (1954, 1964).

9. See note 7.

10. Strictly speaking what they will agree on is that this correlation holds in the class of actual cases of this kind. Modal generalizations might prove contentious.

11. A related criticism has sometimes been made of advocates of backward causation in contemporary physics. For example, in discussions of an advanced action interpretation of quantum mechanics advocated by Costa de Beauregard, Bernard d'Espagnat appears to move from an acknowledgment of the "apparently irreplaceable role of man-centred concepts in the very definition of the causality concept" to the view that there is little of any novelty or promise in the claim that quantum mechanics reveals backward causation. See d'Espagnat (1989a), pp. 229–31, and (1989b), pp. 144–45. I think there is a middle path here, which d'Espagnat misses, but clearly we need to tread carefully. More on the application of these ideas to quantum mechanics in chapter 9.

12. I am ignoring for these purposes the non-Dummettian strategies for avoiding the bilking argument; see note 7.

13. Even here we might in principle avoid advanced action by rejecting correlations toward the future, rather than accepting correlations toward the past. This seems an unpalatable option, however. Without correlations after interactions, for one thing, measurements are likely to prove impossible.

14. Of course, it is important to have shown that the more basic terms employed in chapter 5 do not themselves rest on something which turns out to be perspectival, such as the asymmetry of counterfactual dependence.

15. Here's a more formal way to put this. μInnocence amounts to a T-asymmetric constraint on the structure of the phase space of a physical system—for example, to the requirement that the phase space of a system of photons and polarizers have the following property: If there is a phase space trajectory in which a given photon has some state ϕ at time t, immediately before it encounters a polarizer with setting S_1, then there are other trajectories in which only the setting of the polarizer (or strictly, the setting plus its history, whatever that may be) is different—i.e., in which the

photon has the same state ϕ at *t*, even though the polarizer setting is, say, S_2. This principle is intuitively plausible, despite the fact that the T-inverse principle is highly counterintuitive (and incompatible with orthodox quantum theory, for example).

16. Compare: with hindsight we see that we really had no good reason to expect microphysical objects to be colored, but in order to appreciate that this is the case, we first had to appreciate the subjective nature of color concepts.

CHAPTER 8. *The Puzzle of Contemporary Quantum Theory*

1. Schrödinger (1935), quoted by Lockwood (1989), p. 196.

2. These two terms actually refer to different aspects of the view, but aspects which go together in a natural way. If we want to say that the collapse of the state function corresponds to a change in our information about an objectively existing world, then we seem committed to the existence of such further facts or hidden variables. Conversely, if we want to say that there are hidden variables, then it is a natural move to avoid the measurement problem by denying that collapse corresponds to a real physical change—that is, by embracing the ignorance interpretation. However, not all hidden variable theorists take this course: one of the best-known hidden variable theories, that of David Bohm, holds instead that the wave function is physically real, and avoids the measurement problem by denying that collapse ever happens.

3. The fact that Bell's argument does not depend on the assumption of hidden variables first seems to have been pointed out by Eberhard (1977). There are one or two unusual interpretations of quantum mechanics which claim to avoid nonlocality (without advanced action, which is the strategy I am going to recommend in chapter 9). I shall mention one of the most interesting of these views later in this chapter.

4. For more on this argument see Lockwood (1989), pp. 207–9, and Penrose (1989), pp. 286–87.

5. This objection has been urged by Penrose (1989), pp. 354–56, and Lockwood (1989), pp. 209–10, for example.

6. In other words, a photon between the polarizers is considered to have a state such that it would pass any polarizer oriented in the same direction as the earlier one, but not any polarizer oriented in the same direction as the later one. This formulation points to what may be a more basic asymmetry, however. The very notion of the state seems to be "forward looking": it looks to what *would happen in the future,* under certain conditions. It is plausible that this notion embodies a conventional asymmetry, in much the same way as our ordinary use of counterfactual conditionals does so—as we have just seen, after all, the state itself naturally is characterized in counterfactual terms. But this conclusion would be antithetical to the complete description view, which sees the state function as a codification of an objective reality.

7. Aharonov, Bergmann, and Lebowitz (1964), p. B1410; see also Belinfante (1975).

8. These ideas may be formalized as follows. The question is whether for given a quantum mechanical system S, it is possible in general to describe a collection of underlying states $\{u_j\}$ meeting two requirements: (1) that the knowledge that S had a particular underlying state u_j at a time t would suffice to predict the result of any possible measurement which might be made on S at t; and (2) that there be a rule which for each possible quantum state ψ of S, ascribes a probability $\rho(u_j)$ to each u_j, such that the quantum mechanical probability that a measurement of some property P of S will yield a particular value v, when S has state ψ, is the sum of the probabilities $\rho(u_j)$ over the set of those j for which the underlying state u_j guarantees a result v on a measurement of P.

9. Kochen and Specker (1967).

10. Albert (1994), p. 39. This is the best informal account of Bohm's theory. There is also an excellent introduction and discussion in Albert (1992), ch. 7.

11. Bell (1982).

12. The original paper is Bell (1964). Bell's papers on the subject are collected in Bell (1987). The example on which Bell's argument is based is described in Bohm (1951), pp. 614–19.

13. See particularly Mermin (1981, 1985). Readers familiar with Bell's Theorem, by parable or otherwise, should of course feel free to skip.

14. There are a number of concerns that might arise here, but the one that Doppelgänger seems to have found most pressing is this: if we allow space-like influences of this kind, then if it is not to be an arbitrary matter at what time a given influence "arrives," certain inertial frames must be physically distinguished from others, in violation of the spirit of special relativity.

15. Most notably those of Aspect, Dalibard, and Roger (1982).

16. Gribbin (1990). For a response taking issue with Gribbin's characterization of Bell in these terms, see Price (1991d).

17. The most accessible expositions of the GHZ results in the literature are those of Mermin (1990) and Maudlin (1994), pp. 24–28. My account draws on that of Clifton, Pagonis, and Pitowsky (1992).

18. One of the virtues of the presentation of the GHZ argument in Clifton, Pagonis, and Pitowsky (1992) is the care they take to present the argument against locality in terms which don't assume the presence of hidden variables.

19. With one or two rather esoteric exceptions, at any rate; I mention one of these, the so-called many minds view, later in the chapter.

20. Terminology varies here: Bell's Theorem itself is often classified as one of the no hidden variable results.

21. As I shall use the term here, no collapse views are a subset of the complete description interpretations. This qualification is necessary because some hidden variable views—including David Bohm's—treat the wave function as physically real, and deny that it collapses.

22. This was the term used by Everett (1957), to whom this tradition in quantum mechanics is due. For a good exposition of this view, emphasizing the desirability of stressing its perspectival character, see Lockwood (1989), chs. 12–13.

23. See particularly Healey (1984) and (1989), §6.3, and Hughes (1989), chs. 9–10.

24. Healey (1984), p. 593, and Hughes (1989), p. 293, both point out that there is a problem here, but don't explore the connections with traditional moves in the theory of probability.

25. See Lockwood (1989), pp. 230–32, for a view of this kind.

26. Ironically, some of the proponents of the no collapse view have argued that in virtue of its unique metaphysical view, it is especially well placed to explicate the notion of probability in quantum mechanics. The argument—see DeWitt (1970), Jammer (1974), ch. 11.6—seems to me to involve exactly the same fallacy as would be involved in an attempt to derive the probabilities of coin tosses from the tree structure described in this example.

27. See Lewis (1980) and Mellor (1971) for views of this kind.

28. Albert (1992), p. 131; the many minds view is first proposed in Albert and Loewer (1988).

29. Albert (1992), p. 130.

30. Albert (1992), p. 131.

31. The best survey I know of these ideas is that by Dieter Zeh in Giuline et al. (forthcoming).

32. Zurek (1991), p. 44.

33. Zurek (1991), p. 44.

34. Zurek (1991), p. 44.

35. The words are those of John Baez (private communication, 8 August 1994). Baez himself favors the no collapse view, but my point is that this understanding of the decoherence program—the correct one, in my view—makes it quite compatible with other interpretations of quantum theory.

36. Is there a problem with the asymmetry of the decoherence view? I think that as in Burbury's case, we should point out that it is puzzling that external influences don't "come at haphazard" in both temporal directions. But it is not clear that the decoherence view really need deny this—i.e., not clear that it need say that the number of branches is really decreasing toward the past. How could we tell, since we are only aware of one branch? For more on these issues, see Zeh (1992).

CHAPTER 9. *The Case for Advanced Action*

1. In Davies and Brown (1986), pp. 48–49. The irony actually runs deeper than this, for as we have seen, Bell's Theorem seems to undercut Einstein's strongest argument in favor of his view that there is more to reality than quantum mechanics describes.

2. Davies and Brown (1986), p. 47.

3. Bell (1987), p. 154. Note that under the common past hypothesis it is largely a terminological matter whether we take the factors λ to include the postulated common cause, as Bell's phrasing here suggests, or whether instead we restrict λ to the hidden state of the quantum system, so that both λ and the measurement

settings are effects of a common cause distinct from them both. I shall continue to use the latter convention.

4. Nitpickers might object that the twins are not microscopic systems, so how could μInnocence possibly apply to them? But it is our parable, and we can put the micro–macro boundary wherever we like. Persistent nitpickers are welcome to do this the hard way, using quantum mechanics itself.

5. See note 22.

6. Bell et al. (1985)

7. As we noted in chapter 5, the fact that the bilking argument depends on an assumption of this kind was pointed out by Michael Dummett (1964). Later we shall see that quantum mechanics is tailor-made to exploit Dummett's loophole.

8. 9·1 is just an illustration of the general strategy, of course. In a properly developed theory, something like this would no doubt emerge as a consequence of more basic principles. But our present interest is simply in showing that the general strategy is much more promising than almost everybody has assumed.

9. That is, it avoids *direct* action at a distance, by resolving it into components which lie within light cones. In claiming that advanced action avoids nonlocality, I mean of course that it avoids primitive spacelike influences, which do not resolve in this way into influences which are individually local in the sense allowed by special relativity.

10. There is an up-to-date analysis in Butterfield (1994), and a survey of the field in Maudlin (1994), chs. 4–6.

11. Bell again, in Davies and Brown (1986), p. 50.

12. Bell (1987), p. 110.

13. To the extent that causation itself is regarded as a physically respectable notion, at any rate. Other views are possible, of course, but then the objection to backward causation will not be specifically that it threatens classical realism.

14. Clifton, Pagonis, and Pitowsky (1992), p. 117.

15. Clifton, Pagonis, and Pitowsky (1992), p. 114.

16. Perhaps I should emphasize again that it is *primitive* nonlocality that advanced action promises to avoid; see note 8.

17. Putnam (1979).

18. Putnam (1979), p. 138.

19. Putnam (1979), pp. 138–39.

20. Putnam (1979), pp. 140–41.

21. Wheeler (1978), p. 41.

22. A brief guide to some of these exceptions: the earliest and certainly the most prolific advocate of an advanced action interpretation is O. Costa de Beauregard, whose papers on the topic date back to 1953. His more recent papers include (1977), (1979), and (1985). (Costa de Beauregard's views are discussed by d'Espagnat in the works mentioned in ch. 7, n. 11.) The most highly developed version of the interpretation is perhaps that of Cramer (1986, 1988), on which more in a moment. In the former paper (pp. 684–85) Cramer compares his view to earlier advanced action interpretations. Other advocates of advanced action in quantum mechanics

include Davidon (1976), Rietdijk (1978), Schulman (1986, 1996), and Sutherland (1983). See also Price (1984).

I touched on Cramer's approach in the final section of chapter 3. As I explained there, Cramer takes his motivation from the Wheeler-Feynman Absorber Theory, in a way which seems to me to be questionable, given the difficulties I identified for that theory. Motivation aside, however, how does Cramer's approach to quantum mechanics differ from mine? The big difference is that Cramer takes the wave function to be real, and tries to restore time symmetry by adding another wave function, analogous to Wheeler and Feynman's advanced waves in the radiation case. My view, in contrast, seeks to restore symmetry at the level of hidden variables. I rely on the fact that the asymmetry of the state function is unproblematic, if it is simply an incomplete description. Thus I think Cramer misses the true potential of advanced action, which is to restore something like Einstein's vision of quantum theory.

CHAPTER 10. *Overview*

1. Perhaps the difficulty here isn't immediately obvious, but think about what you would really be wishing for if you wished that you were a different person; that is, if you wished not merely to look at the world through that person's eyes, so to speak—to occupy their social role—but actually to *be* her or him. Suppose you wanted to be Woody Allen, for example. Who would have to look at the world through Woody Allen's spectacles in order for your wish to be granted? Woody Allen himself, obviously, since otherwise it would just be the weaker case, in which a different person stood in his shoes. However, this means that what you wished for is just that Woody Allen himself be Woody Allen, which isn't what you thought you had in mind.

Bibliography

Aharonov, Y., Bergmann, P. G. and Lebowitz, J. L. 1964: "Time Symmetry in the Quantum Process of Measurement," *Physical Review B,* **134,** 1410–16.

Albert, D. Z. 1992: *Quantum Mechanics and Experience,* Cambridge, Mass.: Harvard University Press.

————. 1994: "Bohm's Alternative to Quantum Mechanics," *Scientific American,* **270:5,** 32–39.

Albert, D. Z. and Loewer, B. 1988: "Interpreting the Many Worlds Interpretation," *Synthese,* 77, 195–213.

Arntzenius, F. 1990: "Physics and Common Causes," *Synthese,* **82,** 77–96.

Aspect, A., Dalibard, J. and Roger, G. 1982: "Experimental Test of Bell's Inequalities Using Time-Varying Analyzers," *Physical Review Letters,* **49,** 1804–7.

Augustine, Saint. 1912: *St. Augustine's "Confessions" with an English translation by William Watts (1631), Volume II,* Rouse, W. H. D. (ed.), The Loeb Classical Library, Cambridge, Mass.: Harvard University Press.

Beck, A. (Trans.), 1989: *The Collected Papers of Albert Einstein. Volume 2, The Swiss Years: Writings 1900–1909,* Princeton, N. J.: Princeton University Press.

Belinfante, F. J. 1975: *Measurements and Time Reversal in Objective Quantum Theory,* Oxford: Pergamon Press.

Bell, J. S. 1964: "On the Einstein-Podolsky-Rosen Paradox," *Physics,* **1,** 195–200; reprinted in Bell (1987).

————. 1981: "Bertlmann's Socks and the Nature of Reality," *Journal de Physique,* **42,** C2-41–C2-62; reprinted in Bell (1987).

————. 1982: "On the Impossible Pilot Wave," *Foundations of Physics,* **12,** 989–99; reprinted in Bell (1987).

————. 1987: *Speakable and Unspeakable in Quantum Mechanics: Collected Papers on Quantum Philosophy,* Cambridge University Press.

Bell, J. S., Clauser, J., Horne, M. and Shimony, A. 1985: "An Exchange on Local Beables," *Dialectica,* **39,** 85–110.

Blackburn, S. 1993: *Essays in Quasi-Realism,* New York: Oxford University Press.

Bohm, D. 1951: *Quantum Theory,* Englewood Cliffs, N. J.: Prentice Hall.

————. 1952: "A Suggested Interpretation of Quantum Theory in Terms of Hidden Variables," *Physical Review,* **85,** 166–93.

Boltzmann, L. 1877: "Über die Beziehung eines allgemeine mechanischen Satzes zum zweiten Hauptsatze der Wärmetheorie," *Sitzungsberichte, K. Akademie der*

Wissenschaften in Wien, Math.-Naturwiss., 75, 67–73; English translation ("On the Relation of a General Mechanical Theorem to the Second Law of Thermodynamics") in Brush (1966), 188–93.

———. 1895a: "On Certain Questions of the Theory of Gases," *Nature*, 51, 413–15.

———. 1895b: "On the Minimum Theorem in the Theory of Gases," *Nature*, 52, 221.

———. 1964: *Lectures on Gas Theory*, Berkeley: University of California Press.

Bronstein, M. and Landau, L. 1933: *Soviet Physics*, 4, 113.

Brush, S. 1966: *Kinetic Theory. Volume 2: Irreversible Processes*, Oxford: Pergamon Press.

———. 1976: *The Kind of Motion We Call Heat. Book 2: Statistical Physics and Irreversible Processes*, Amsterdam: North Holland.

———. 1983: *Statistical Physics and the Atomic Theory of Matter*, Princeton, N. J.: Princeton University Press.

Burbury, S. H. 1894: "Boltzmann's Minimum Function," *Nature*, 51, 78.

———. 1895: "Boltzmann's Minimum Function," *Nature*, 51, 320.

Burman, R. 1970: *Observatory*, 90, 240–49.

———. 1971: *Observatory*, 91, 141–45.

———. 1972: *Observatory*, 92, 128–35.

———. 1975: *Physics Letters A*, 53, 17–18.

Burtt, E. A. 1954: *The Metaphysical Foundations of Modern Physical Science*, Garden City, N.Y.: Doubleday Anchor Books.

Butterfield, J. 1994: "Stochastic Einstein Nonlocality and Outcome Dependence," in Prawitz, D. and Westerståhl, D. (eds.), *LMPS91, Selected Papers from the Uppsala Congress*, Dordrecht: Kluwer, 385-424.

Clausius, R. 1865: *Annalen der Physik, Series 2*, 125, 426.

———. 1867: "On Several Convenient Forms of the Fundamental Equations of the Mechanical Theory of Heat," in Hirst, T. A. (ed.), *The Mechanical Theory of Heat*, London: Van Voorst.

Clifton, R., Pagonis, C. and Pitowsky, I. 1992: "Relativity, Quantum Mechanics and EPR," in Hull, D., Forbes, M. and Okruhlik, K. (eds.), *PSA 1992, Volume 1*, Chicago: Philosophy of Science Association, 114–28.

Collingwood, R. G. 1940: *An Essay in Metaphysics*, Oxford University Press.

Costa de Beauregard, O. 1953: "Méchanique Quantique," *Académie des Sciences*, 1632.

———. 1977: "Time Symmetry and the Einstein Paradox," *Il Nuovo Cimento*, 42B, 41–64.

———. 1979: "Time Symmetry and the Einstein Paradox—II," *Il Nuovo Cimento*, 51B, 267–79.

———. 1985: "On Some Frequent but Controversial Statements concerning the Einstein-Podolsky-Rosen Correlations," *Foundations of Physics*, 15, 871–87.

Cramer, J. G. 1980: "Generalized Absorber Theory and the Einstein-Podolsky-Rosen Paradox," *Physical Review D*, 22, 362–76.

————. 1983: "The Arrow of Electromagnetic Time and the Generalized Absorber Theory," *Foundations of Physics,* **13**, 887–902.

————. 1986: "The Transactional Interpretation of Quantum Mechanics," *Reviews of Modern Physics,* **58**, 647–87.

————. 1988: "An Overview of the Transactional Interpretation of Quantum Mechanics," *International Journal of Theoretical Physics,* **27**, 227–36.

Csonka, P. 1969: "Advanced Effects in Particle Physics," *Physical Review,* **180**, 1266–81.

Culverwell, E., 1890a: "Note on Boltzmann's Kinetic Theory of Gases, and on Sir W. Thomson's Address to Section A, British Association, 1884," *Philosophical Magazine,* **30**, 95–99.

————. 1890b: "Possibility of Irreversible Molecular Motions," *Report of the British Association for the Advancement of Science,* **60**, 744.

————. 1894: "Dr. Watson's Proof of Boltzmann's Theorem on Permanence of Distributions," *Nature,* **50**, 617.

Davidon, W. C. 1976: "Quantum Physics of Single Systems," *Il Nuovo Cimento,* **36B**, 34–40.

Davies, P. C. W. 1970: *Proceedings of the Cambridge Philosophical Society,* **68**, 751–64.

————. 1971: *Journal of Physics,* **A4**, 836–45.

————. 1972: *Journal of Physics,* **A5**, 1025–36.

————. 1974: *The Physics of Time Asymmetry,* London: Surrey University Press.

————. 1977: *Space and Time in the Modern Universe,* Cambridge University Press.

————. 1983: "Inflation and Time Asymmetry in the Universe," *Nature,* **301**, 398–400.

————. 1995: *About Time: Einstein's Unfinished Revolution,* London: Viking.

Davies, P. C. W. and Brown, J. R. (eds.), 1986: *The Ghost in the Atom,* Cambridge University Press.

Davies, P. C. W. and Twamley, J. 1993: "Time-symmetric Cosmology and the Opacity of the Future Light Cone," *Classical and Quantum Gravity,* **10**, 931.

d'Espagnat, B. 1989a: *Reality and the Physicist,* Cambridge University Press.

————. 1989b: "Nonseparability and the Tentative Descriptions of Reality," in Schommers, W. (ed.), *Quantum Theory and Pictures of Reality,* Berlin: Springer-Verlag, 89–168.

DeWitt, B. S. 1970: "Quantum Mechanics and Reality," *Physics Today,* **23**, 30–35.

Dummett, M. A. E. 1954: "Can an Effect Precede Its Cause?" *Proceedings of the Aristotelian Society, Supplementary Volume,* **38**, 27–44.

————. 1964: "Bringing about the Past," *Philosophical Review,* **73**, 338–59.

Eberhard, P. H. 1977: "Bell's Theorem without Hidden Variables," *Il Nuovo Cimento,* **38B**, 75–80.

Einstein, A. and Ritz, W. 1909: "Zum gegenwärtigen Stand des Strahlungsproblems" ("On the Present Status of the Radiation Problem"), *Physikalische Zeitschrift,* **10**, 323–24; English translation in Beck(1989), 376.

Einstein, A., Podolsky, B. and Rosen, N. 1935: "Can Quantum-Mechanical Description of Physical Reality Be Considered Complete?" *Physical Review,* 47, 777–80.

Everett, H. 1957: "'Relative State' Formulation of Quantum Mechanics," *Reviews of Modern Physics,* 29, 452–64.

Gasking, D. 1955: "Causation and Recipes," *Mind,* 64, 479–87.

Gell-Mann, M. and Hartle, J. 1994: "Time Symmetry and Asymmetry in Quantum Mechanics and Quantum Cosmology," in Halliwell, Perez-Mercader, and Zurek (1994), pp. 311–45.

Giuline, D., Joos, E., Kiefer, C., Kupsch, J., Stamatescu, I.-O. and Zeh, H. D. Forthcoming: *Decoherence and the Appearance of a Classical World in Quantum Theory,* Berlin: Springer-Verlag.

Gleick, J. 1992: *Genius: Richard Feynman and Modern Physics,* London: Little, Brown and Company.

Gold, T. 1962: "The Arrow of Time," *American Journal of Physics,* 30, 403–10.

———. (ed.), 1967: *The Nature of Time,* Ithaca: Cornell University Press.

Gribbin, J. 1990: "The Man Who Proved Einstein Was Wrong," *New Scientist,* 24 November 1990, 33–35.

Grünbaum, A. 1973: *Philosophical Problems of Space and Time,* Dordrecht: Reidel.

Halliwell, J. 1994: "Quantum Cosmology and Time Asymmetry," in Halliwell, Perez-Mercader, and Zurek (1994), pp. 369–89.

Halliwell, J. and Hawking, S. W. 1985: "Origin of Structure in the Universe," *Physical Review D,* 31, 1777–91.

Halliwell, J., Perez-Mercader, J. and Zurek, W. (eds.), 1994: *Physical Origins of Time Asymmetry,* Cambridge University Press.

Hausman, D. M. 1986: "Causation and Experimentation," *American Philosophical Quarterly,* 23, 143–54.

Hawking, S. W. 1985: "Arrow of Time in Cosmology," *Physical Review D,* 33, 2489–95.

———. 1988: *A Brief History of Time,* London: Bantam.

———. 1994: "The No Boundary Condition and the Arrow of Time," in Halliwell, Perez-Mercader, and Zurek (1994), pp. 346–57.

Healey, R. A. 1984: "How Many Worlds?" *Noûs,* 18, 591–616.

———. 1989: *The Philosophy of Quantum Mechanics: An Interactive Interpretation,* Cambridge University Press.

Hogarth, J. E. 1962: "Cosmological Considerations of the Absorber Theory of Radiation," *Proceedings of the Royal Society,* A267, 365–83.

Horwich, P. 1975: "On Some Alleged Paradoxes of Time Travel," *Journal of Philosophy,* 72, 432–44.

———. 1987: *Asymmetries in Time,* Cambridge, Mass.: MIT Press.

Hoyle, F. and Narlikar, J. V. 1974: *Action at a Distance in Physics and Cosmology,* San Fransisco: J. V. Freeman.

Hughes, R. I. G. 1989: *The Structure and Interpretation of Quantum Mechanics,* Cambridge, Mass.: Harvard University Press.

Hunter, J. 1788: *Philosophical Transactions of the Royal Society of London,* **78**, 53.

Jackson, F. 1977: "A Causal Theory of Counterfactuals," *Australasian Journal of Philosophy,* **55**, 3–21.

Jammer, M. 1974: *The Philosophy of Quantum Mechanics,* New York: Wiley.

Johnston, M. 1989: "Dispositional Theories of Value," *Proceedings of the Aristotelian Society Supplementary Volume,* **63**, 139–74.

————. 1993: "Objectivity Refigured: Pragmatism Without Verificationism," in Wright, C. and Haldane, J. H. (eds.), *Reality, Representation and Projection,* Oxford University Press, 85–130.

Kochen, S. and Specker, E. P. 1967: "The Problem of Hidden Variables in Quantum Mechanics," *Journal of Mathematics and Mechanics,* **17**, 59–87.

Lewis, D. 1976: "The Paradoxes of Time Travel," *American Philosophical Quarterly,* **13**, 145–52.

————. 1980: "A Subjectivist's Guide to Objective Chance," in Jeffrey, R. C. (ed.), *Studies in Inductive Logic and Probability,* Berkeley: University of California Press; reprinted in Lewis (1986a), 83–113.

————. 1986a: *Philosophical Papers, Volume II,* Oxford University Press.

————. 1986b: *The Plurality of Worlds,* Oxford: Blackwell.

Linde, A. 1987: "Inflation and Quantum Cosmology," in Hawking, S. W. and Israel, W. (eds.), *Three Hundred Years of Gravitation,* Cambridge University Press, 604–30.

Lockwood, M. 1989: *Mind, Brain and the Quantum: The Compound "I,"* Oxford: Basil Blackwell.

Lyons, G. W. 1992: "Complex Solutions for the Scalar Field Model of the Universe," *Physical Review D,* **46**, 1546.

Maxwell, J. 1867: "On the Dynamical Theory of Gases," *Philosophical Transactions of the Royal Society of London,* **157**, 49–88.

Maudlin, T. 1994: *Quantum Non-locality and Relativity,* Oxford: Basil Blackwell.

Mellor, D. H. 1971: *The Matter of Chance,* Cambridge University Press.

————. 1981: *Real Time,* Cambridge University Press.

Menzies, P. and Price, H. 1993: "Causation as a Secondary Quality," *British Journal for the Philosophy of Science,* **44**, 187–203.

Mermin, N. D. 1981: "Quantum Mysteries for Anyone," *Journal of Philosophy,* **78**, 397–408.

————. 1985: "Is the Moon There When Nobody Looks? Reality and Quantum Theory," *Physics Today,* **38:4**, 38–47.

————. 1990: "What's Wrong with These Elements of Reality?" *Physics Today,* **43:6**, 9–11.

Nagel, T. 1986: *The View from Nowhere,* New York: Oxford University Press.

Narlikar, J. V. 1962: *Proceedings of the Royal Society,* **A270**, 553–61.

Newton, I. 1952: *Opticks,* 4th London ed. of 1730, Dover.

Page, D. N. 1983: "Inflation Does Not Explain Time Asymmetry," *Nature,* **304**, 39–41.

Papineau, D. 1985: "Causal Asymmetry," *The British Journal for the Philosophy of Science,* **36**, 273–89.

Partridge, R. B. 1973: "Absorber Theory of Radiation and the Future of the Universe," *Nature,* **244**, 263–65.

Penrose, O. and Percival, I. C. 1962: "The Direction of Time," *Proceedings of the Physical Society,* **79**, 605–16.

Penrose, R. 1979: "Singularities and Time-Asymmetry," in Hawking, S. W. and Israel, W. (eds.), *General Relativity: An Einstein Centenary,* Cambridge University Press, 581–638.

———. 1989: *The Emperor's New Mind,* Oxford University Press.

Pettit, P. 1991: "Realism and Response-Dependence," *Mind,* **100**, 587–626.

Popper, K. 1956a: "The Arrow of Time," *Nature,* **177**, 538.

———. 1956b: "Reply to Schlegel," *Nature,* **178**, 382.

Price, H. 1984: "The Philosophy and Physics of Affecting the Past," *Synthese,* **16**, 299–323.

———. 1989: "A Point on the Arrow of Time," *Nature,* **340**, 181–2.

———. 1991a: Review of Denbigh, K. and Denbigh, J., *Entropy in Relation to Incomplete Knowlege,* and Zeh, H. D., *The Physical Basis of the Direction of Time, British Journal for the Philosophy of Science,* **42**, 111–44.

———. 1991b: "Agency and Probabilistic Causality," *British Journal for the Philosophy of Science,* **42**, 157–76.

———. 1991c: "The Asymmetry of Radiation: Reinterpreting the Wheeler-Feynman Argument," *Foundations of Physics,* **21**, 959–75.

———. 1991d: "Saving free will," *New Scientist,* 12 January 1991, 55–56.

———. 1991e: "Two Paths to Pragmatism," in Menzies, P. (ed.), *Working Papers in Philosophy,* Canberra: Research School of Social Sciences, ANU, 46–82.

———. 1992a: "Agency and Causal Asymmetry," *Mind,* **101**, 501–20.

———. 1992b: "Metaphysical Pluralism," *Journal of Philosophy,* **79**, 387–409.

———. 1993: "The Direction of Causation: Ramsey's Ultimate Contingency," in Hull, D., Forbes, M. and Okruhlik, K. (eds.), *PSA 1992, Volume 2,* 253–267.

———. 1994: "A Neglected Route to Realism about Quantum Mechanics," *Mind,* **103**, 303–36.

———. 1995: "Cosmology, Time's Arrow and That Old Double Standard," in Savitt, S. (ed.), *Time's Arrows Today,* Cambridge University Press, 66–94.

Putnam, H. 1979: "A Philosopher Looks at Quantum Mechanics," *Philosophical Papers, Volume I,* 2nd ed., Cambridge University Press, 130–58.

Ramsey, F. P. 1978: "General Propositions and Causality," in Mellor, D. H. (ed.), *Foundations: Essays in Philosophy, Logic, Mathematics and Economics,* London: Routledge and Kegan Paul, 133–51.

Reichenbach, H. 1956: *The Direction of Time,* Berkeley: University of California Press.

Rietdijk, C. W. 1978: "Proof of a Retroactive Influence," *Foundations of Physics,* **8**, 615–28.

Russell, B. 1963: *Mysticism and Logic,* London: George Allen & Unwin.

Sachs, R. G. 1987: *The Physics of Time Reversal,* University of Chicago Press.

Schrödinger, E. 1935: *Die Naturwissenschaften,* **23**, 807–12, 824–28, 844–49.

Schulman, L. S. 1986: "Deterministic Quantum Evolution through Modification of the Hypothesis of Statistical Mechanics," *Journal of Statistical Physics,* **42**, 689.

———. 1996: *Time's Arrows and Quantum Measurement,* Cambridge University Press.

Sikkima, A. E. and Israel, W. 1991: "Black-hole Mergers and Mass Inflation in a Bouncing Universe," *Nature,* **349**, 45–47.

Sklar, L. 1992: *Philosophy of Physics,* Oxford University Press.

Smart, J. J. C. 1955: "Spatialising Time," *Mind,* **64**, 239–41.

———. 1963: *Philosophy and Scientific Realism,* London: Routledge & Kegan Paul.

Stephenson, L. M. 1978: "Clarification of an Apparent Asymmetry in Electromagnetic Theory," *Foundations of Physics,* **8**, 921–26.

Sutherland, R. I. 1983: "Bell's Theorem and Backwards-in-Time Causality," *International Journal of Theoretical Physics,* **22**, 377–84.

Thom, P. 1974: "Time-Travel and Non-Fatal Suicide," *Philosophical Studies,* **27**, 211–16.

Thomson, W. 1852: "On a Universal Tendency," *Proceedings of the Royal Society of Edinburgh,* **3**, 139.

Tooley, M. 1987: *Causation,* Oxford: Clarendon Press.

van Fraassen, B. 1989: *Laws and Symmetry,* Oxford: Clarendon Press.

von Weizsäcker, C. 1939: "Der zweite Haupsatz und der Unterschied von der Vergangenheit und Zukunft," *Annalen der Physik (5 Folge),* **36**, 275–83.

———. 1980: *The Unity of Nature,* New York: Farrar Straus Giroux.

von Wright, G. H. 1975: *Causality and Determinism,* New York: Columbia University Press.

Wheeler, J. 1978: "The 'Past' and the 'Delayed Choice' Double-Slit Experiment," in Marlow, A. (ed.), *Mathematical Foundations of Quantum Theory,* New York: Academic Press, 9–48.

Wheeler, J. A. and Feynman, R. P. 1945: "Interaction with the Absorber as the Mechanism of Radiation," *Reviews of Modern Physics,* **17**, 157–181.

Williams, D. C. 1951: "The Myth of Passage," *Journal of Philosophy,* **48**, 457–72.

Wright, C. 1993: "Realism: The Contemporary Debate—W(h)ither Now?" in Wright, C. and Haldane, J. H. (eds.), *Reality, Representation and Projection,* Oxford University Press, 63–84.

Zeh, H. D. 1989: *The Physical Basis of the Direction of Time,* Berlin: Springer-Verlag.

———. 1992: *The Physical Basis of the Direction of Time,* 2nd ed., Berlin: Springer-Verlag.

Zermelo, E. 1896: "Ueber Einen Satze der Dynamik und die mechanische Wärmtheorie," *Annalen der Physik, Series 3,* **57**, 485; English trans. in Brush (1966), 208.

Zurek, W. H. 1991: "Decoherence and the Transition from Quantum to Classical," *Physics Today,* **44:10**, 36–44.

Index